Modern Molecular
Orbital Theory
for
Organic Chemists

WESTON T. BORDEN

University of Washington

Modern Molecular
Orbital Theory
for
Organic Chemists

PRENTICE-HALL, INC.

Englewood Cliffs, New Jersey

Library of Congress Cataloging in Publication Data

BORDEN, WESTON T. (date)
  Modern molecular orbital theory for organic chemists.

  Includes bibliographical references.
  1. Molecular orbitals. 2. Chemistry, Organic.
I. Title.
QD461.B66   547'.1'22   74–17245
ISBN   0-13-595983-7

© 1975 by Prentice-Hall, Inc.
Englewood Cliffs, New Jersey

10 9 8 7 6 5 4 3 2 1

Printed in the United States of America

PRENTICE-HALL INTERNATIONAL, INC., *London*
PRENTICE-HALL OF AUSTRALIA, PTY. LTD., *Sydney*
PRENTICE-HALL OF CANADA, LTD., *Toronto*
PRENTICE-HALL OF INDIA PRIVATE LIMITED, *New Delhi*
PRENTICE-HALL OF JAPAN, INC., *Tokyo*

To
Buffy Pittipaw Wamsuta
and
Flower Rumblina Eleganza

Contents

Preface

This book evolved from my lecture notes for Chemistry 137, a one-semester course in theoretical organic chemistry for first–year graduate students and advanced seniors that I gave for three years at Harvard University. The goal of the course was to teach the students sufficient theory so that they could read the literature critically and carry out intelligent calculations of their own.

Modern molecular orbital theory has progressed far beyond simple Hückel π electron calculations, the mastery of which was probably sufficient in the early sixties for the organic chemist desirous of keeping abreast of the theoretical organic literature. The development of the Woodward-Hoffmann rules for pericyclic reactions has sparked an explosion of interest in theory among organic chemists and in organic systems among theoreticians. The result is that today, unless a chemist is familiar with such topics as perturbation theory, CNDO/2 calculations, and group theory, a large segment of the important literature of organic chemistry must remain obscure to him. Chemistry 137 was designed to familiarize students in organic chemistry with just such arcane-sounding subjects, rather than giving them a terminal course in quantum mechanics, concerned with more traditional topics like the harmonic oscillator, rigid rotor, and hydrogen atom.

At the time that I began teaching Chemistry 137, I had no intention of writing a textbook. Nevertheless, I, or to be more precise, the students, soon discovered that none of the available books was suitable for first–year graduate students in organic chemistry. They found that textbooks containing theory more sophisticated than the Hückel method become too mathematical too early. This is not to denigrate the mathematical ability of Harvard students studying physical chemistry. However, the graduate students who enrolled in Chemistry 137 were almost all organically inclined; and it has been my observation that organic chemists, myself included, tend to understand physical pictures much better than mathematical abstractions. Therefore, especially in the early part of the course, the presentation relied much more heavily on physical intuition than on mathematical rigor.[1] Only in the second half, when the students had developed a good physical "feeling" for molecular orbital theory, did the course become more mathematical.

This text, which is the result of expansion and frequent revision of the xeroxed lecture notes that I felt compelled to hand out to the students, may also be roughly divided into two parts. The first presents a one-electron, Hückel picture of the behavior of electrons in molecules, particularly organic molecules. Little mathematics beyond high school algebra and a smattering of calculus is required in this portion of the text. No knowledge of quantum mechanics is assumed, but some previous exposure in an introductory physical chemistry course would certainly be helpful. The use of symmetry, where it exists, is emphasized throughout the book; but in the first part the nonmixing rule for wavefunctions of different symmetries and the relationship between the LCAO-MO coefficients of symmetry equivalent atoms are justified by intuitive physical arguments. In the second half of the book these intuitive arguments are replaced by mathematical derivations of the properties of Hermitian operators and the introduction of group theory. The necessity of employing correctly antisymmetrized many-electron wavefunctions is developed, and the consequences of including electron repulsion in the Hamiltonian forms the central theme of the second half of the text. For this portion of the book some previous contact with matrices and determinants would be useful but is not mandatory, since the mathematics necessary for their manipulation is outlined.

An important pedagogical feature of this text is that the organic molecules, discussed from the point of view of one-electron theory in the first half of the text, are re-examined in the second with the electron repulsion operator included in the Hamiltonian. Treating the same subject material twice not only allows the student to concentrate his attention on the necessarily more complex mathematical manipulations required with correctly antisymmetrized many-electron wavefunctions, but it also

---

[1] Some physical chemists will undoubtedly feel that presenting qualitative physical pictures before rigorous mathematical derivations is putting the cart before the horse. Such people may be appalled at the absence from the first chapter of detailed treatments of such topics as the variational principle, the eigenvalue problem and secular determinants. These topics were covered later in the course and consequently appear in the second half of the book, because I found that their early introduction only proved an unnecessary and unwelcome burden to most of the organic chemists for whom the course was given and for whom this book was written.

enables him to appreciate the improvement obtained in the description of these molecules when a more complete theory is employed. Throughout the first half of the text the failings of Hückel theory are noted, and in the second half it is shown how these deficiencies are overcome by the inclusion of electron repulsion in the Hamiltonian. I found that this process encouraged the students to struggle with the additional mathematical complexity, encountered in the second half of the book, and helped motivate them to go beyond the mathematically and physically simple one-electron picture with which they had become comfortable. Nevertheless, because the first half of the book forms a complete unit, it could be used by itself as the text for part of an advanced organic chemistry course that emphasized MO theory. I have, in fact, used it for such a course at the University of Washington.

The last chapter of the book, which is concerned with all-valence electron calculations, differs from those preceding it, in that it tends to be more of a survey. The three major types of calculations—extended Hückel, __NDO, and ab initio—are examined, their approximations and consequent limitations discussed, and their successes and failures briefly reviewed. References are provided to the major reviews and books, concerned with all-valence electron calculations, that have recently appeared.

At the end of each of the previous chapters just one or two suggestions are given for further reading that I feel will be of genuine use to the student, interested in pursuing or seeing an alternate presentation of a particular subject. Since different theoreticians use different notations, reading original papers in theory can prove difficult, time consuming and, hence, discouraging for organic chemists. For this reason I have attempted to avoid citing the original literature, preferring instead to direct the reader to other textbooks and reviews. I must, therefore, apologize to all those theoreticians whose contributions I have discussed without giving them proper credit by referencing their original papers. Leading references to the original literature can be found in the sources that I have cited as suggestions for further reading.

The reader will find that throughout the text are interspersed a large number of problems; in fact, over 170 of them appear. It has been my own personal experience, as well, I believe, as that of most others, that in trying to learn theory, there is no substitute for doing problems. In addition, the problems presented here often amplify or expand on points that are only mentioned briefly in the text. Worked examples are also included in abundance; indeed, the key chapters, 3 and 7, are organized by the molecules whose electronic structures are worked out. Nevertheless, the reader will find that only by doing the problems at the end of a portion of text can he be assured of having mastered the material presented in that section. Moreover, such drill should implant the material in that section firmly enough in his mind so that he will probably find it unnecessary to refer back to it as he proceeds through the text. I hope that the large number of problems included in this book will make it useful as a self-instruction manual for organic chemists who wish to teach themselves modern MO theory.

   I want to thank the students who took Chemistry 137 for their patience in trying to decipher the more cryptic sections of earlier versions of the lecture notes and for calling these sections to my attention. The style of the final text greatly benefited from my severest critic, the former Marcia Robbins, who mercilessly truncated the run-on sentences that proliferated in the earliest versions of the manuscript. Finally, I wish to thank Mary Glenn Goldman for typing the entire manuscript and remaining consistently cheerful throughout the numerous revisions of it.

WESTON THATCHER BORDEN

Seattle, Washington

1

Hückel Theory—
A Semiquantitative
Model of Bonding

The organic chemist is accustomed to thinking of the properties of molecules as deriving from the nature of and interplay between the component functional groups. It may seem natural to him, therefore, to attempt a theoretical description of the chemical bond in terms of the characteristics of the constituent atoms and their interactions. This is the approach that we shall begin to explore in this chapter.

The Hydrogen Molecule

Let us start with the simplest neutral molecule, H_2. We will seek a solution for the electrons in the hydrogen molecule of the Schrödinger equation

$$\mathcal{H}\Psi = E\Psi \qquad (1)$$

The Hamiltonian operator, $\mathcal{H}$, contains, in addition to terms for the electrostatic potentials present in a molecule or atom, a differential operator which gives the kinetic energy of the electrons. Therefore, (1) is really a differential equation; and when it is solved, it yields a set of wavefunctions, Ψ, and the particular value of the energy, E, associated with each. The Schrödinger equation can only be solved in closed form for very simple systems like the hydrogen atom. In the hydrogen atom the solutions, Ψ, of the Schrödinger equation are wavefunctions for a single electron. These are commonly referred to as *orbitals*. The square of one of these wavefunctions, or orbitals, when evaluated at a given point in space gives the probability at that point of finding an electron which occupies the orbital. If an orbital contains an electron, the total probability of finding the electron somewhere in space is one. It is customary, therefore, to normalize orbitals so that

$$\int \Psi^2 \, d\tau = 1 \tag{2}$$

where the integral is taken over all space. All wavefunctions are generally normalized in this way.

We may consider that from solution of the Schrödinger equation we know the orbitals and their energies for two hydrogen atoms at infinite separation. We now wish to find the orbitals and the associated orbital energies of the hydrogen molecule. Direct solution of the Schrödinger equation for this system is impossible, so let us try to find an approximate solution by considering what happens as the separated hydrogen atoms are brought together and allowed to interact. Since the hydrogen $1s$ atomic orbitals are by far the lowest in energy, only they will be filled; and we can confine our attention to them. Because of the wavelike properties of the orbitals, their interaction will depend on their relative signs, as shown schematically in Figure 1.1. If they are brought together in phase, they will add in the region of interaction. If they are out of phase, they will tend to cancel out each other. Therefore, from a linear combination of two atomic orbitals (AO's) we can expect two possible molecular orbitals (MO's). One will have a large electron density in the region between the nuclei where the AO's add. The other, which has a node (a plane at which the wavefunction goes to zero because of the change in sign) between the nuclei, will tend to exclude electron density from this region. Since a buildup of electron density between the nuclei is associated with bond formation, the

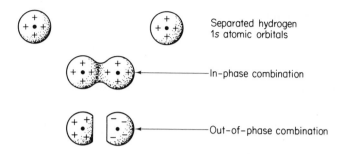

Separated hydrogen
1s atomic orbitals

—In-phase combination

—Out-of-phase combination

Linear Combinations of Hydrogen 1s Atomic Orbitals

Figure 1.1

in-phase combination of AO's must give a *bonding MO*, whose energy is expected to be lower than that of the out-of-phase combination. As we shall see, the out-of-phase combination corresponds, in fact, to an *antibonding MO*.

Problem 1.1.   Show that if two AO's, ϕ_1 and ϕ_2, are added, the resulting probability density is greater than $\phi_1^2 + \phi_2^2$, the result that would be obtained if the densities, rather than the wavefunctions, added.

Problem 1.2.   Suppose ϕ_1 and ϕ_2 are properly normalized so $\int \phi_1^2 \, d\tau = \int \phi_2^2 \, d\tau = 1$. Is the MO wavefunction $1/\sqrt{2}(\phi_1 + \phi_2)$ properly normalized?

We will now show mathematically that construction of an MO from a linear combination of AO's (LCAO) gives the results anticipated on the basis of qualitative considerations. We write the LCAO-MO as

$$\psi = c_1\phi_1 + c_2\phi_2 \qquad (3)$$

where ψ[1] is the MO that we are attempting to construct, ϕ_1 and ϕ_2 are the 1s AO's of the two hydrogen atoms, and c_1 and c_2 are the coefficients we must find to determine ψ. On substituting (3) into the Schrödinger equa-

---

[1] Throughout the remainder of this book the symbol ψ will be used for MO wavefunctions, ϕ for AO's, and capital Ψ for the complete wavefunction for all the electrons in a molecule.

tion (1),[2] we obtain

$$\mathcal{H}c_1\phi_1 + \mathcal{H}c_2\phi_2 = c_1 E\phi_1 + c_2 E\phi_2 \tag{4}$$

If we multiply both sides of (4) by ϕ_1 and integrate, we get

$$c_1 \int \phi_1 \mathcal{H}\phi_1 \, d\tau + c_2 \int \phi_1 \mathcal{H}\phi_2 \, d\tau = c_1 E \int \phi_1^2 \, d\tau + c_2 E \int \phi_1\phi_2 \, d\tau \tag{5}$$

since the c's are constants and may be removed from the integration. Similarly, multiplication by ϕ_2 followed by integration gives

$$c_1 \int \phi_2 \mathcal{H}\phi_1 \, d\tau + c_2 \int \phi_2 \mathcal{H}\phi_2 \, d\tau = c_1 E \int \phi_2\phi_1 \, d\tau + c_2 E \int \phi_2^2 \, d\tau \tag{6}$$

We will now try to identify some of these integrals with quantities that we can hope to calculate or measure. Let us begin by considering the integral $\int \phi_1 \mathcal{H}'\phi_1 \, d\tau$ for an isolated hydrogen atom, where $\mathcal{H}'$ is the Hamiltonian operator for the isolated atom. Since ϕ_1 is a $1s$ hydrogen AO, it must solve

$$\mathcal{H}'\phi_1 = \alpha'\phi_1 \tag{7}$$

where α' is the energy of an electron in a $1s$ orbital on an isolated hydrogen atom. We assume that ϕ_1 is normalized; and since α' is a constant that can be removed from the integral,

$$\int \phi_1 \mathcal{H}'\phi_1 \, d\tau = \int \phi_1 \alpha'\phi_1 \, d\tau = \alpha' \int \phi_1^2 \, d\tau = \alpha' \tag{8}$$

In the hydrogen molecule the corresponding integral, $\int \phi_1 \mathcal{H}\phi_1 \, d\tau$, must be equal to the energy of an electron localized on atom 1. We shall denote

---

[2] This step requires a certain amount of faith; for although solution of the Schrödinger equation gives the total electronic wavefunction, Ψ, and its associated energy, we have no guarantee that $\mathcal{H}\psi = E\psi$ can be used to find a molecular orbital—the wavefunction for an individual electron. Of course, in a molecule like H_2^+, or in a neutral hydrogen atom, there is only one electron, and this problem does not arise, since $\Psi = \psi$ (or ϕ in the hydrogen atom). For systems which do contain more than one electron, we shall prove in a subsequent chapter that an equation of the same form as (1) can, in fact, be used to find orbitals.

the energy of an electron localized on atom 1 in the molecule by the symbol α_1. Similarly, $\int \phi_2 \mathcal{K} \phi_2 \, d\tau = \alpha_2$ is the energy of an electron localized on atom 2. Finally, the quantity $\int \phi_1 \mathcal{K} \phi_2 \, d\tau$ may be identified with the energy of the electron density in the region between the nuclei, since $\phi_1 \phi_2$ represents the electron density in the region where ϕ_1 and ϕ_2 overlap. The energy of this electron density is given the symbol β, which is often referred to as the resonance integral. The summation over space of the overlap density, $\int \phi_1 \phi_2 \, d\tau$, is given the symbol S and called the overlap integral. We shall discuss these quantities more fully in a later section of this chapter; for the moment let us assume that we know how to compute them. Equations (5) and (6) are now replaced by

$$c_1 \alpha_1 + c_2 \beta = c_1 E + c_2 S E \tag{9}$$

$$c_1 \beta + c_2 \alpha_2 = c_1 S E + c_2 E \tag{10}$$

These are two equations in three unknowns, so we can solve for the energy E and the ratio of the coefficients c_1 and c_2. The magnitudes of the coefficients can then be computed from the normalization condition (2). Solving (9) for c_2, we obtain

$$c_2 = \frac{c_1(E - \alpha_1)}{\beta - SE} \tag{11}$$

Rearranging (10) and substituting for c_2 gives

$$c_1(\beta - SE) = c_1 \frac{(E - \alpha_1)(E - \alpha_2)}{(\beta - SE)} \tag{12}$$

which has the trivial solution $c_1 = 0$. A nontrivial solution exists only if the equation

$$(\beta - SE)^2 = (E - \alpha_1)(E - \alpha_2) \tag{13}$$

is satisfied. This means that only certain values of the energy are allowed. Remembering that we are dealing with a hydrogen molecule so that $\alpha_1 = \alpha_2$, we obtain

$$(\beta - SE)^2 = (E - \alpha)^2 \tag{14}$$

which has two solutions

$$(\beta - SE) = \pm(E - \alpha) \tag{15}$$

Since α and β both represent the energy of an electron in an attractive potential, they are both negative numbers. Thus, the solution with the plus sign gives the lowest energy

$$E = \frac{\alpha + \beta}{1 + S} \tag{16}$$

which corresponds to the most stable orbital. According to the Pauli exclusion principle, this bonding orbital can accommodate both electrons in the hydrogen molecule, provided their spins are antiparallel.

Problem 1.3. Up to this point we have no guarantee that (16) represents the lowest possible energy that can be obtained using an LCAO-MO wavefunction. Show that, in fact, it does represent the lowest energy by writing the Schrödinger equation in the integrated form $\int \psi H\psi \, d\tau = E \int \psi^2 \, d\tau$, assuming ψ to have the form of Equation (3), and minimizing E with respect to variations in c_1 and c_2. You should obtain Equations (9) and (10).

The MO of energy $E = (\alpha + \beta)/(1 + S)$ can be found by substitution of Equation (16) into (11). The result is

$$c_2 = c_1 \frac{\dfrac{(\alpha + \beta) - \alpha(1 + S)}{1 + S}}{\dfrac{\beta(1 + S) - S(\alpha + \beta)}{1 + S}} = \frac{\beta - \alpha S}{\beta - \alpha S} c_1 = c_1 \tag{17}$$

This then represents the in-phase combination of the AO's, as we would expect from the fact that this orbital has the lowest energy. The wavefunction corresponding to the antibonding MO of energy

$$E = \frac{\alpha - \beta}{1 - S} \tag{18}$$

has $c_1 = -c_2$ and represents the out-of-phase combination.

Problem 1.4.   The bonding MO has the form $\psi = c(\phi_1 + \phi_2)$ and the antibonding MO $\psi = c(\phi_1 - \phi_2)$. Show that for normalization the correct value of c is $1/\sqrt{2 + 2S}$ for the former and $1/\sqrt{2 - 2S}$ for the latter.

Use of Molecular Symmetry

We have now obtained the bonding and antibonding MO's for H_2 and found their associated energies. However, we could have found the same solutions much more easily by making use of the molecular symmetry. The two nuclei in a hydrogen molecule are indistinguishable; therefore, in any MO the number of electrons on one atom must be the same as that on the other. It is easy to show (Problem 1.5) that the number of electrons q, localized on atom A in a normalized MO ψ_i is

$$q_A^i = c_{Ai}^2 \tag{19}$$

for each electron in the MO. Therefore, in H_2 symmetry requires that $c_1^2 = c_2^2$ which allows two possibilities, $c_1 = c_2$ and $c_1 = -c_2$. So from symmetry alone it is possible to determine the two MO's for H_2; and by use of the integrated form of the Schrödinger equation, their energies can be found.

Problem 1.5.   Demonstrate that Equation (19) is correct.

Problem 1.6.   Determine the energy of the two MO's derived from symmetry by using the integrated form of the Schrödinger equation.

With regard to the application of symmetry, the hydrogen molecule is typical. We shall find that where symmetry exists and can be utilized, the simplification of computational problems is often considerable. Moreover, as we shall see, the application of symmetry can often provide deeper insight into a problem than the performance of a calculation can. This is due to the fact that the results obtained from the application of symmetry, where it exists, are necessarily correct, while the results of a calculation may be dependent on the approximations inherent in the computational method employed.

The Parameters α, β, and S

Three parameters α, β, and S, are involved in the expressions that we have derived for the energy of the orbitals in the hydrogen molecule. Of these S is the most straightforward to compute. Given two orbitals and the separation between them, its evaluation is trivial; however, it is clear that if we could somehow get rid of the overlap integral, the algebra in our calculations would be greatly simplified.

The calculation of α might also appear to be straightforward, since it is tempting to identify this quantity with α', the energy of one electron in an isolated hydrogen atom. However, this would not be strictly correct; for, if we are considering the neutral molecule containing two electrons, expansion of the MO wavefunction shows that in the molecule the probability that both electrons will simultaneously be found localized on one atom is the same as the probability that they will be found one on each atom. Since the electrons repel each other electrostatically, the former electron distribution is of higher energy than the latter. In this simple theory, where we do not explicitly take into account electron repulsion, we must provide for its implicit inclusion in some average way by our choice of parameters. We will eventually see that the failure to explicitly account for electron repulsion in the Hamiltonian is the major shortcoming of this model of chemical bonding, introduced by Hückel in 1931, and which is generally referred to as Hückel theory.

Problem 1.7. In this simple theory the wavefunction for the ground state of a hydrogen molecule may be written $\Psi = \psi(1)\psi(2)$, where ψ is the bonding MO and the numbers in parentheses refer to its occupation by electron 1 and electron 2. Compute the probability density, $\Psi^2 = \psi^2(1)\psi^2(2)$, for the ground state, showing that it is equally probable to find the electrons simultaneously localized on the same atom as on different atoms. [Hint: Substitute $(1/\sqrt{2 + 2S})(\phi_A + \phi_B)$ for ψ and show that terms like $\phi_A^2(1)\phi_A^2(2)$ and $\phi_B^2(1)\phi_B^2(2)$ have the same coefficients as the terms $\phi_A^2(1)\phi_B^2(2)$ and $\phi_B^2(1)\phi_A^2(2)$.]

The ionization potential, I, of a neutral atom is the energy required to remove its least tightly bound electron, and this quantity may be roughly equated with the negative of the energy of that electron when it resides on the atom. The electron affinity, $\mathcal{A}$, is the corresponding quantity for the

8

uninegative ion. From the equal probability of finding both electrons on different atoms or on the same atom in the H_2 molecule, we might expect that α, the energy of an electron localized on one atom, should be set equal to the average of $-I$ and $-\mathcal{Q}$. The following argument further justifies this value for α. Theoretically, the change in energy on transferring two electrons from a uninegative atom A to a unipositive atom B is $2\alpha_B - 2\alpha_A$. Experimentally, we would measure this change as $\mathcal{Q}_A - I_B$ for the first electron plus $I_A - \mathcal{Q}_B$ for the second electron. Equating the two expressions for the change in energy,

$$2\alpha_B - 2\alpha_A = \mathcal{Q}_A - I_B + I_A - \mathcal{Q}_B \qquad (20)$$

Since α_A can only depend on quantities related to atom A, we arrive at

$$\alpha = -\tfrac{1}{2}(I + \mathcal{Q}) \qquad (21)$$

This expression, except for the factor of $-\tfrac{1}{2}$, is just Mulliken's definition of the electronegativity of an atom.[3]

Having arrived at a more or less satisfactory definition of α, we can now proceed to examine the physical significance of β more closely. β represents the energy of the overlap charge distribution given by $\phi_A\phi_B$. Mulliken has suggested as a useful approximation to an overlap charge distribution

$$\phi_A\phi_B = \tfrac{1}{2}S(\phi_A^2 + \phi_B^2) \qquad (22)$$

However, this cannot be used directly to compute β, for we are not interested in the quantity

$$\int \mathcal{H}\phi_A\phi_B \, d\tau = \tfrac{1}{2}S\left(\int \mathcal{H}\phi_A^2 \, d\tau + \int \mathcal{H}\phi_B^2 \, d\tau\right) \qquad (23)$$

but want to obtain $\beta = \int \phi_A\mathcal{H}\phi_B \, d\tau$. These two quantities are not the same, for we must remember that $\mathcal{H}$ is a differential operator and the commutative law therefore does not hold. However, a form for β suggested

---

[3] There is a close correlation between Mulliken's theoretically based definition of electronegativity and Pauling's values, which are derived empirically from bond energies. In the section on heteronuclear diatomics in the next chapter, we shall see why this correspondence exists.

by (23) is not unreasonable. We might expect the total energy of the overlap density, $\phi_A\phi_B$, to be proportional to the total amount of it, S, and to the average of the energy of an electron localized on A and on B. We might then try a modified form of (23), first used by Wolfsberg and Helmholtz,

$$\beta = \tfrac{1}{2}KS(\alpha_A + \alpha_B) \tag{24}$$

It is necessary to assign K a value greater than one in order to obtain reasonable results with the Wolfsberg-Helmholtz expression for β. Let us examine the consequences of setting $K = 1.0$. With this value of K $\beta = S\alpha$, and the total energy of the hydrogen molecule would then be[4]

$$E(H_2) = 2\left(\frac{\alpha + \beta}{1 + S}\right) = 2\left(\frac{\alpha + S\alpha}{1 + S}\right) = 2\alpha \tag{25}$$

This is greater than the energy of two isolated hydrogen atoms, since $2\alpha = -I - \mathcal{Q} > -2I$. Clearly, the energy of the electrons in the overlap region must be considerably lower than the average energy of an electron localized on one of the bonded atoms, or H_2 molecules would not exist. That is why a value of $K > 1$ is required.

Now let us write the energy of the bonding orbital of the hydrogen molecule using the Wolfsberg-Helmholtz formula for β with $K > 1$.

$$E = \frac{\alpha + \beta}{1 + S} = \frac{\alpha + KS\alpha}{1 + S} \tag{26}$$

This can be rewritten as

$$E = \frac{\alpha(1 + S)}{1 + S} + \frac{K'S\alpha}{1 + S} \tag{27}$$

where $K' = K - 1$. If we now define

$$\beta' = \frac{K'S\alpha}{1 + S} \tag{28}$$

---

[4] This assumes that the total energy is the sum of orbital energies. This assumption is quite incorrect, since it counts electron repulsions twice. If writing the total energy as a sum of orbital energies works at all, it is due to a fortuitous cancellation of other electrostatic terms that we have neglected. We will discuss this failing of Hückel theory, in which total energies are written as sums of orbital energies, much more fully in later chapters.

then (27) becomes

$$E = \alpha + \beta' \qquad (29)$$

All the dependence of the energy on S is now contained in β'. This means that we can now omit explicitly carrying S through our calculations, provided we remember the functional dependence of β' on S. This redefinition of the resonance integral, which allows *formal* neglect of S [S can be set equal to 0 in Equations (9)–(18)], is very convenient; for instance, the normalization constant for both H_2 wavefunctions becomes simply $1/\sqrt{2}$. Because of the many simplifications that it allows, we shall employ exclusively this redefined β (i.e., β'), which formally allows us to set $S = 0$.[5] Therefore, having no more need for the old β, we can, for notational simplicity, drop the prime from the redefined resonance integral. We should remember, however, that the redefined β is no longer the energy of the electron density in the overlap region, but is now equal to $S/(1 + S)$, the fraction of each electron found in the overlap region, times the *difference* in energy between an electron in the overlap region and one on an atom in the molecule.

Problem 1.8.   Show that the corresponding quantity β'' in the expression for the energy of the antibonding orbital $E = \alpha - \beta''$ must be defined as $\beta'' = K'S\alpha/(1 - S)$. Show that this implies: (a) "Antibonding orbitals are more antibonding than bonding orbitals are bonding" and (b) A "spectroscopic" β, measured by half the difference in energy between the bonding and antibonding MO's, will not be the same as the one that appears in the expression for the energy of the bonding MO.

Problem 1.9.   For a molecule made up of several identical atoms the general expression for the energy of an orbital is

$$E = \alpha + \frac{m}{1 + mS}(\beta - \alpha S) \qquad (30)$$

where m is a constant that depends on how the atoms are joined (their connectivity) and on the particular MO in question. Rewrite (30), using the Wolfsberg-Helmholtz formula, so it has the form it would if S were equal to zero.

$$E = \alpha + m\beta' \qquad (31)$$

[5] In general this cannot be completely justified theoretically. (See Problem 1.9.)

Show that β' depends on m and that thus β' is not strictly a parameter which can be transferred from molecule to molecule or even orbital to orbital (See Problem 1.8). Show, nevertheless, that if m and S are small numbers, the dependence of β' on m is small.

Let us now try to relate the parameter $K'S\alpha/(1 + S)$, which we henceforth denote by the symbol β, to quantities that we can measure experimentally. Let us calculate the dissociation energy of the hydrogen molecule, $D(H_2)$, which is equal to the energy of the separate atoms minus the energy of the molecule. Writing the total energy of the molecule as the sum of the orbital energy for each electron[6] and substituting from (21) for α,

$$E(H_2) = 2\alpha + 2\beta = -I - \alpha + 2\beta \qquad (32)$$

The energy of two hydrogen atoms at infinite separation is just $-2I$. Thus, for $D(H_2)$ we calculate

$$D(H_2) = -2I + I + \alpha - 2\beta = -I + \alpha - 2\beta \qquad (33)$$

We can now express β in terms of experimentally measurable quantities. Rearranging (33)

$$-2\beta = D(H_2) + I - \alpha \qquad (34)$$

It can be shown (See Problem 1.10) that $I - \alpha$ represents the increase in electron–electron repulsion associated with bond formation in Hückel theory. Because of the shortcomings of the Hückel method in dealing with electron repulsion, particularly in expressing total energies as simply the sums of orbital energies, (34) does not give a quantitatively correct expression for β in the H_2 molecule. Nevertheless, it is correct in indicating that 2β is really greater *in magnitude* than the bond dissociation energy of the hydrogen molecule by an amount equal to the actual increase in electron repulsion on bond formation.

Problem 1.10.   Consider two identical atoms at infinite separation, one bearing a unit negative charge, the other a unit positive charge. If an electron is transferred from the negative to the positive ion, giving two neutral atoms, the only change in total energy in this process arises from the re-

---

[6] See Footnote 4.

moval of the repulsion energy, γ_{aa}, between two electrons on the same atom. What is the change in energy in terms of I and α? Use this result to show that the difference between -2β and the dissociation energy of H_2 is, as given by (34), equal to the repulsion energy between two electrons on the same atom. Compare this result with that of Problem 1.7, and show that the coefficient of γ_{aa} in (34) should be $\frac{1}{2}$. The error of a factor of two in (34) results from expressing the total energy as the sum of orbital energies, which counts the electron repulsion twice, once for each electron.

The Origin of Bond Formation in the Hydrogen Molecule[7]

The hydrogen molecule exists because its energy is lower than that of two separate hydrogen atoms. In the previous section we saw that the reason the hydrogen molecule has a positive dissociation energy is that 2β is greater in magnitude than the increase in electron repulsion concomitant with bond formation. Recalling that our redefined β is proportional to the energy difference between an electron in the overlap region and one on an atom in the molecule, we conclude that the hydrogen molecule exists because the energy of an electron in the region between the two nuclei is considerably lower than that of an electron residing near one hydrogen nucleus. It is tempting to ascribe the lower energy of an electron in the overlap region to a decrease in its potential energy, because in this region an electron will feel most strongly the simultaneous attraction of both nuclei. Not only does this picture appeal to our classical (as opposed to quantum mechanical) physical intuition; but it also appears to be supported by the following argument, which does indeed show that bond formation in the hydrogen molecule occurs because of a lowering of the potential energy of the electrons. The argument relies on a well-known theorem of physics, called the virial theorem, which states that at (and only at) infinite separation or the equilibrium bond distance, the kinetic (T), potential (V), and total (E) energy of two hydrogen atoms obey the following relationship

$$E = \tfrac{1}{2}V = -T \tag{35}$$

---

[7] The material contained in this section is not essential to an understanding of the rest of the book.

Since the virial theorem applies at infinite separation and at the equilibrium bond length, we can apply it to the change in energy (ΔE) on going from isolated hydrogen atoms to the hydrogen molecule.

$$\Delta E = \tfrac{1}{2}\Delta V = -\Delta T \qquad (36)$$

Thus, since we know the total energy decreases on going from the atoms to the molecule, the potential energy must also decrease—twice as much in fact—while the kinetic energy increases. Since our LCAO-MO model shows that the decrease in the total energy is due to a decrease in the energy of an electron in the overlap region, it is concluded from the virial theorem that in this region the potential energy of an electron must be decreased. Therefore, this argument seems to show that the classical electrostatic picture is correct; but in fact, both the argument and the picture are in error.

It is considerably easier mathematically to carry out detailed quantum mechanical calculations on H_2^+ than on H_2, since in the molecular ion there is only one electron and hence no electron repulsion. The origin of bond formation in both molecules is, however, the same; and detailed calculations on H_2^+ show that the potential energy of an electron in the overlap region is actually greater than that of an electron on an atom in the molecule. While it is true that in the overlap region an electron simultaneously feels a similar attraction from both nuclei, the magnitude of the attractive potential is not as large as that near one nucleus.

The quantity which does show an algebraic decrease in the overlap region is the kinetic energy of an electron, and this is a purely quantum mechanical, rather than a classical effect. It can be understood qualitatively on the basis of the Heisenberg uncertainty principle, which states

$$\Delta p_x \Delta x = \hbar \qquad (37)$$

where Δp_x is the uncertainty in momentum along the x axis, Δx is the uncertainty in position along this axis, and $\hbar$ is Planck's constant divided by 2π. When there is a large uncertainty in the position of an electron, not only is the uncertainty in its momentum small, but the momentum itself tends to be small. When there is less uncertainty in the position of an electron, that is, if an electron is localized in a smaller region of space, its momentum and consequently its kinetic energy increase. A manifestation of this consequence of the uncertainty principle is the familiar quantum mechanical result for a particle in a box that, as the size of the box increases, the kinetic energy of the particle decreases and vice versa.

Now in the region between the two nuclei of the H_2^+ molecule, the MO wavefunction is relatively flat, since it is the sum of two exponentials (which represent the radial parts of the two $1s$ AO's), one of which increases in moving from left to right along the internuclear axis and the other of which decreases. In fact, in this region it can be shown (Problem 1.11) that the density of the overlap charge is constant along the internuclear axis. Thus, since the probability of finding an electron at any one point in the overlap region is similar to finding it at any other, there must be a large uncertainty in the exact position of an electron in the overlap region. Since the uncertainty in position is large, the momentum and hence the kinetic energy of an electron in the overlap region must be small. It is just this reduction in kinetic energy, compared to that of an electron on an atom (where the wavefunction decreases exponentially so that the positional uncertainty is small), which gives rise to bond formation. The only role of the potential in the region between the two nuclei is to make it possible for an electron to lower its kinetic energy by moving into this region without suffering too large an increase in its potential energy. Nevertheless, it is to be emphasized that the potential energy of an electron in the overlap region is higher than that of an electron on one of the atoms.

Problem 1.11.   Write the radial part of the $1s$ wavefunctions for two hydrogen atoms as e^{-r_A} and e^{-r_B}, where $r_A = 0$ at nucleus A and $r_B = 0$ at nucleus B. What is the function which describes the overlap density? Show mathematically that this function is constant along the internuclear axis in the region between the nuclei. [Hint: What is the value of $r_A + r_B$ here?]

Problem 1.12.   What changes do you expect in the potential and kinetic energy of an electron in the antibonding MO of H_2^+ as the overlap between the atoms increases? Give a brief physical interpretation of why these changes occur.

If, as the LCAO–MO model tells us, the existence of bonding in the hydrogen molecule (or molecular ion) depends on the lowered energy in the overlap region, and if in this region it is the kinetic and not the potential energy of an electron that is lowered, then bonding must result from a decrease in the kinetic energy of the electrons. This conclusion seems to be in direct contradiction with that derived from the virial theorem, which asserts [Equation (36)] that the kinetic energy actually increases on bond formation and that bonding is a consequence of the twofold larger decrease in the potential energy. How can we reconcile these two results? The answer is that the virial theorem applies to the hydrogen molecule and not necessarily to our LCAO–MO model for it.

An LCAO–MO wavefunction can, however, be made to satisfy the virial theorem, provided allowance is made for possible changes in the AO's of atoms in constructing the MO's. If the size of the AO's is allowed to vary by including scale factors in the exponentials which represent their radial parts, and if the energy of the molecule is minimized with respect to variations in the scale factors, the resulting wavefunction can be shown to satisfy automatically the virial theorem. The total energy of an MO wavefunction thus optimized is, of course, lower than that of one which uses the same AO's for both the separated atoms and the molecule. However, much larger than the change in the total energy that results from using optimized AO's are the changes in the kinetic and potential energy of the molecule. These changes come about because the optimized AO's are contracted in size with respect to those in the free atoms. Since an electron in a contracted AO is held closer to the nucleus, its potential energy is decreased; and because an electron in such an orbital is more localized, its kinetic energy is increased. These changes that occur on varying the size of the AO's in the molecule to minimize the energy (and so obtain a wavefunction which satisfies the virial theorem) are *intra*atomic in the sense that even with such a wavefunction the kinetic energy of an electron is still lower and its potential energy higher in the overlap region than on one of the atoms in the molecule. Thus, if we consider a molecule to be created from atoms whose AO's are already optimized for bond formation, we again find that the decisive factor in making the energy of the molecule lower than that of the isolated atoms is the lowering of the kinetic energy of the electrons in the overlap region.

Summary

We constructed a model for bond formation in H_2 by using an LCAO–MO approach and showed that the wave nature of AO's gives rise to two MO's, $\psi = 1/\sqrt{2}(\phi_1 + \phi_2)$ and $\psi = 1/\sqrt{2}(\phi_1 - \phi_2)$, depending on whether the AO's combine in or out of phase. The energies of the MO's are respectively $E = \alpha + \beta$ and $E = \alpha - \beta$, when β is defined so as to allow the formal neglect of S. The magnitude of α was identified with the average of the ionization potential I and electron affinity $\mathcal{Q}$ of the atom. β, when defined so that S can be ignored, was seen to be proportional to the amount by which the energy of an electron is lower in the overlap region than on an atom. Finally, this energy lowering in the overlap region, which is responsible for bond formation, was traced to a decrease in the kinetic energy of an electron in this region.

FURTHER READING

J. C. SLATER, *Quantum Theory of Molecules and Solids*, McGraw-Hill Book Co., New York, 1963. Chapters 1–4 give an excellent description in quantitative terms of the theory of bonding in the H_2^+ and H_2 molecules. The interested reader can use the equations for H_2^+ in Chapter 2 to demonstrate that the kinetic energy of an electron in the overlap region is, in fact, lower and the potential energy higher than that of an electron on one of the atoms in the molecule. [Hint: In comparing the relative energy contributions of the atoms and the overlap region, remember the relative probabilities of finding an electron in each.]

W. KUTZELNIGG, *Angew. Chem. Int. Edit.* **12**, 546 (1973) presents a fine qualitative discussion of bonding.

2

Application of
Hückel Theory to
Di and Triatomic Molecules

We have used the known stability of the hydrogen molecule to develop a simple, if not naive, picture of chemical bonding. In this chapter we will try to make use of this theory to predict the stability of some diatomic and the geometry of some triatomic molecules.

Homonuclear Diatomics

Application of symmetry to the interaction of two identical orbitals on two identical atoms assures us that the correct MO wavefunctions are

$$\psi = \frac{1}{\sqrt{2}}(\phi_1 + \phi_2) \quad \text{and} \quad \psi = \frac{1}{\sqrt{2}}(\phi_1 - \phi_2) \tag{1}$$

with the plus sign giving a bonding MO of energy $E = \alpha + \beta$ and the minus, the antibonding MO of energy $E = \alpha - \beta$. Of course, different sets of AO's will require different values of α and β.

It is worth noting at this point that if bonding and antibonding MO's are equally occupied, net antibonding—that is, net instability with respect to dissociation into two atoms—will result. This arises from the fact, demonstrated in Problem 1.8, that the magnitude of β in an antibonding MO is greater than that in a bonding MO.

Since we know the MO's in homonuclear diatomics from symmetry, let us begin with a description of the bonding in this class of molecules. In H_2 both electrons can be placed in a bonding MO, and we know the molecule to be stable. There are in He_2, however, four electrons. Since the Pauli exclusion principle allows only two electrons to be placed in the bonding MO, the remaining two must go into the antibonding orbital. Hence, we predict no stability for He_2, and none is observed.

In the first row diatomics both the bonding and antibonding MO's derived from the $1s$ AO's are filled, so that any contribution that the electrons in these MO's make to bonding will be small. In fact, when we begin to investigate how to write correct wavefunctions for many-electron systems, we will be able to show that when both bonding and antibonding MO's are equally filled, the electrons may be considered to be approximately localized in the constituent AO's. Consequently, in the first row diatomics the $1s$ electrons may be thought of as being largely localized, two on each atom. We will, therefore, only examine the interactions between the $2s$ and $2p$ atomic orbitals in these molecules. The bonding and antibonding MO's which result from the linear combinations of these AO's are pictured in Figure 2.1.

It is useful to classify the MO's by whether they change sign on rotation by $180°$ about an axis through the nuclei. This is just the familiar σ and π designation of bonds, and it depends only on the nature and orientation of the AO's used to form the bond. It also proves useful to further

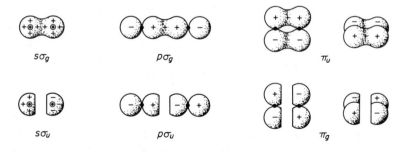

$s\sigma_g$ $p\sigma_g$ π_u

$s\sigma_u$ $p\sigma_u$ π_g

Molecular Orbitals for First Row Diatomics

Figure 2.1

classify the MO's as to their behavior on inversion through the center of the molecule, which, for a given σ or π orbital, depends only on whether it is bonding or antibonding. Note, however, that while a bonding σ orbital is symmetric to inversion (denoted g for the German *gerade*, even), a bonding π orbital changes sign and is thus antisymmetric (symbolized by the label u for *ungerade*). The symmetry labels are shown below the orbitals in Figure 2.1.

We must now consider whether there will be any mixing between these MO's, arising from bonding interactions between their constituent AO's. For instance, will mixing between $s\sigma_u$ and $p\sigma_g$ be brought about through the interaction of a $2s$ AO on one atom and a $2p$ AO, with a σ type orientation on the other? The reader should convince himself that the bonding and antibonding interactions between the two MO's are equal and will cancel. In general we will find that there is no interaction between orbitals of different symmetry, so that the symmetry labels simplify the problem by allowing us to pick out immediately which orbitals will interact.

Problem 2.1.   Show mathematically that the $s\sigma_u$ and the $p\sigma_g$ MO's have no net interaction. Do the same for $s\sigma_g$ and π_u.

Problem 2.2.   Show that the two π_u MO's do not interact, despite the fact that they both have the same symmetry label.

The noninteraction of the π_u MO's is due to the fact that it is really the pair of MO's, taken together, which are assigned the symmetry label. A pair of MO's like these, which by symmetry have the same energy and do not interact, is said to be degenerate. They are given a symmetry designation, which is also termed degenerate, because it is understood to apply to the pair of orbitals taken together. One reason for this is that if two noninteracting orbitals are indistinguishable by symmetry, a completely equivalent set of new orbitals can be formed by taking orthogonal linear combinations of the original ones. Therefore, orbitals of this type cannot be treated separately, but only in pairs. We will have much more to say about degenerate orbitals later.

Problem 2.3.   Show that the MO's formed by adding and subtracting the two $2\pi_g$ MO's are equivalent to the originals, differing only by a $45°$ change in orientation.

The only bonding MO's between which we expect any interaction are the $s\sigma_g$ and the $p\sigma_g$, because they are of the same σ_g symmetry. Although an s and a p orbital on the same atom do not interact, because of their different "local" symmetry, bonding between an s and an appropriately oriented p orbital on different atoms can occur, as inspection of Figure 2.1 shows. It should be possible, therefore, to construct two new MO's from linear combinations of the $s\sigma_g$ and $p\sigma_g$ orbitals. As in the case of the hydrogen molecule, the in-phase combination will build up electron density between the nuclei, while the out-of-phase combination will move electron density away from this region. There is no reason to suppose, however, that the mixing of the MO's will be equal, as it was by symmetry for the AO's in the hydrogen molecule. Let us see why.

From symmetry the interacting MO's can immediately be written as

$$s\sigma_g = \frac{1}{\sqrt{2}}(s_1 + s_2) \quad \text{and} \quad p\sigma_g = \frac{1}{\sqrt{2}}(p_1 + p_2) \qquad (2)$$

We are looking for a new MO of the form

$$\psi = c_1 s\sigma_g + c_2 p\sigma_g \qquad (3)$$

which we know from Problem 1.3 will minimize the energy.

We proceed as with the solution for the hydrogen molecule. Substituting (3) into the Schrödinger equation, multiplying by s_1 and integrating, we obtain

$$c_1\left[\int s_1 \mathcal{H} s_1 \, d\tau + \int s_1 \mathcal{H} s_2 \, d\tau\right] + c_2\left[\int s_1 \mathcal{H} p_1 \, d\tau + \int s_1 \mathcal{H} p_2 \, d\tau\right]$$
$$= Ec_1\left[\int s_1^2 \, d\tau + \int s_1 s_2 \, d\tau\right] + Ec_2\left[\int s_1 p_1 \, d\tau + \int s_1 p_2 \, d\tau\right] \qquad (4)$$

Substituting for the integrals the parameters α_s, α_p, β_s, β_p and β_{sp} gives

$$c_1(\alpha_s + \beta_s) + c_2\beta_{sp} = c_1 E \qquad (5)$$

since atomic symmetry makes $\int s_1 \mathcal{H} p_1 \, d\tau = 0$, and all overlap integrals are formally set equal to zero. The same procedure after multiplication by

p_1 yields

$$c_1\beta_{sp} + c_2(\alpha_p + \beta_p) = c_2 E \qquad (6)$$

Problem 2.4. Show that multiplication by s_2 and p_2 yields equations identical to (5) and (6), respectively.

To emphasize the similarity of this problem to the one that we solved in the first chapter, we can define $\alpha_1 = \alpha_s + \beta_s$ and $\alpha_2 = \alpha_p + \beta_p$, resulting in equations identical to (9) and (10) of Chapter 1, since we assume β_{sp} has been properly defined to allow us to set $S = 0$ in these equations. Now, however, there is no reason why α_1 should equal α_2, as the symmetry of the hydrogen molecule required. Therefore, (13) of Chapter 1 does not have the simple solution $E = \alpha \pm \beta$; and in order to find the two new orbital energies, we must use the quadratic formula, which gives

$$E = \tfrac{1}{2}[\alpha_1 + \alpha_2 \pm \sqrt{(\alpha_1 + \alpha_2)^2 - 4\alpha_1\alpha_2 + 4\beta^2}] \qquad (7)$$

If we now had values for the various parameters required in (7), we could solve for the orbital energies and use them to determine, via (11) of Chapter 1, how the two σ_g orbitals mix. Let us instead make use of (7), to examine two extreme cases in orbital mixing. When $\alpha_1 = \alpha_2$ we get two new orbitals of energy $\alpha + \beta$ and $\alpha - \beta$ corresponding to equal in- and out-of-phase mixing, as we saw for the H_2 molecule. In this case, however, $\alpha_1 < \alpha_2$, since an electron in orbital 1, the MO constructed from the $2s$ AO's, has lower energy. Rewriting (7), using $(\alpha_1 + \alpha_2)^2 - 4\alpha_1\alpha_2 = (\alpha_1 - \alpha_2)^2$,

$$E = \frac{1}{2}\left[\alpha_1 + \alpha_2 \pm (\alpha_2 - \alpha_1)\sqrt{1 + \frac{4\beta^2}{(\alpha_2 - \alpha_1)^2}}\right] \qquad (8)$$

If the difference in the energy of the orbitals, $\Delta\alpha$, is large compared to the energy of their interaction, β, we can use a power series expansion.[1]

$$E = \frac{1}{2}\left[\alpha_1 + \alpha_2 \pm (\alpha_2 - \alpha_1)\left(1 + \frac{1}{2}\frac{4\beta^2}{(\alpha_2 - \alpha_1)^2} + \cdots\right)\right] \qquad (9)$$

[1] The radical $\sqrt{1 + x}$ can be expanded in a power series, $1 + \tfrac{1}{2}x - \tfrac{1}{8}x^2 + \ldots$, when $x < 1$. If $x \ll 1$, it is often sufficiently accurate to retain only the first term containing x.

This equation leads to the two new energies[2]

$$E = \alpha_2 + \frac{\beta^2}{\Delta\alpha} \quad \text{and} \quad E = \alpha_1 - \frac{\beta^2}{\Delta\alpha} \tag{10}$$

where $\Delta\alpha = \alpha_2 - \alpha_1$ and hence is positive.

Because $\Delta\alpha$ is positive, the energy of the lower orbital decreases, and this is accompanied by an equal (when overlap is neglected) increase in energy of the upper. Depending on the relative magnitudes of β and $\Delta\alpha$ we may not always be able to use (10) and may instead have to employ (7) to compute accurately the energies after mixing. Nevertheless, the qualitative result—that, on mixing, the lower of two orbitals further lowers its energy at the expense of the upper—is quite general, and is one we shall often use.

Problem 2.5. Use the results summarized in (10) to find the two new unnormalized MO's from (11) of Chapter 1. Does the in-phase combination correspond to the lowest MO?

Problem 2.6. We shall prove in a later chapter that two different functions which exactly satisfy equations of the Schrödinger type, known as eigenvalue equations, are orthogonal. This means that the integral $\int \psi_1 \psi_2 \, d\tau = 0$. Show that this is the case for the two hydrogen MO's. Assuming that the starting MO's are orthogonal, show that the two wavefunctions you obtained from (11) of Chapter 1 in the previous problem are only approximately orthogonal, so long as $\beta/\Delta\alpha$ has a finite size. Why does orthogonality hold only approximately for these two wavefunctions?

Returning now to the specific problem of mixing the $s\sigma_g$ and $p\sigma_g$ orbitals in a homonuclear diatomic, we see that we will get two new MO's of the form, before normalization,

$$1\sigma_g = s\sigma_g + cp\sigma_g \quad \text{and} \quad 2\sigma_g = s\sigma_g - c'p\sigma_g \tag{11}$$

where c and c' can be obtained from the ratio of the values of c_1 and c_2 which satisfy (5) and (6) for the two permitted values of E.

---

[2] The reader may recognize (10) as the second-order energy term in the treatment of this problem by perturbation theory, which we will discuss in some detail later in this chapter and use extensively throughout this book.

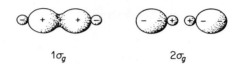

$$1\sigma_g \qquad\qquad\qquad 2\sigma_g$$

Figure 2.2

The MO's are shown in Figure 2.2, and two points appear from the drawings. The first is that the two new MO's still have σ_g symmetry after interaction, and such conservation of symmetry on mixing MO's is to be expected. The second point is that these new MO's have the appearance of hybrid orbitals which, in fact, they are. This can be seen quantitatively by trying to construct a bonding MO from two hybrid AO's on identical atoms. By symmetry, equal amounts of these hybrids will be mixed in-phase to form the bonding MO. Therefore, we can write

$$\psi = s_1 + cp_1 + s_2 + cp_2 \qquad\qquad (12)$$

where c is to be determined so that the hybridization is optimized to give an energy minimum. Were we to carry out the necessary calculation, we would find that c has the same allowed values as in (11); in fact, using the definitions of $s\sigma_g$ and $p\sigma_g$, it is clear that (11) and (12) are identical. In MO theory, therefore, optimum hybridization is automatically achieved by allowing mixing of MO's of the same symmetry. Furthermore, we can see that hybridization is really the response of an atom to bonding; for in general, we can say nothing about the optimum hybridization of an atom until we have examined its interactions with the other atoms in a molecule. This should be clear from the fact that we cannot guess c in (12) without doing a calculation. However, we can be relatively certain that neither $1\sigma_g$ nor $2\sigma_g$ is an sp hybrid, despite the fact that this type of hybrid is expected from elementary valence theory. Such a hybrid would require $c = c' = 1$ in these MO's, which would only be obtained if $s\sigma_g$ and $p\sigma_g$ had the same energy.

Problem 2.7. Sketch the orbitals, $1\sigma_u$ and $2\sigma_u$, that result from the mixing of $s\sigma_u$ with $p\sigma_u$.

We are now ready to order the MO's of a homonuclear diatomic according to energy. We know that in an atom containing many electrons, the $2s$ orbital is lower in energy than the $2p$ orbitals. Therefore, we would expect the $s\sigma_g$ and the $s\sigma_u$ orbitals to lie lowest, especially since their energy will be further lowered by their interaction with, respectively, the $p\sigma_g$ and $p\sigma_u$ MO's. This interaction will raise the energy of these latter MO's. Thus, after mixing with $s\sigma_g$, $p\sigma_g$ will probably lie close to, and perhaps even above, the bonding π_u orbitals. The energy levels of the interacting AO's and the resulting MO's are shown schematically in Figure 2.3.[3]

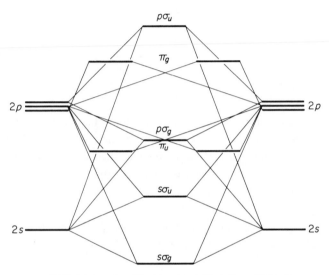

Orbital Interaction Diagram for a First Row Diatomic

Figure 2.3

Table 2.1 shows the order in which the MO's are expected to be filled in the first row diatomics and gives the net number of bonding electrons and predicted stability for each molecule.

Our predictions regarding the expected stability of these diatomics are confirmed experimentally. We can go on to make some further predictions through the use of Hund's rule, which states that when two electrons are placed in a pair of degenerate orbitals like π_u or π_g, the triplet state, in which the electrons occupy separate MO's and have parallel spins,

[3] The MO's have been given the names of the AO's which make the largest contribution to them. However, as indicated in Figure 2.3, the σ MO's contain a mixture of s and p AO's; and to avoid any suggestion to the contrary, they might instead have been given the designations $1\sigma_g$, $1\sigma_u$, $2\sigma_g$, and $2\sigma_u$.

Table 2.1

Filling of the MO's in First Row Homonuclear Diatomics

Molecule	$2s\sigma_g$	$2s\sigma_u$	π_u	$p\sigma_g$	π_g	$p\sigma_u$	Net Bonding Electrons	Stability
Li_2	2						2	yes
Be_2	2	2					0	no
B_2	2	2	2				2	yes
C_2	2	2	4				4	yes
N_2	2	2	4	2			6	yes
O_2	2	2	4	2	2		4	yes
F_2	2	2	4	2	4		2	yes
Ne_2	2	2	4	2	4	2	0	no

is the most stable. The reason is, as we will see in later chapters, that when two electrons have parallel spins, the total electronic wavefunction, Ψ, tends to keep them apart, thus minimizing their electrostatic repulsion. Inspection of Table 2.1 shows that both B_2 and O_2 should have triplet ground states; and this, indeed, is found to be the case.

Problem 2.8. Explain why in C_2 the lowest singlet and triplet states are found to have almost the same energy.

Heteronuclear Diatomics

In a diatomic molecule whose nuclei are different, symmetry cannot be used to find the correct wavefunction. To find the coefficients c_1 and c_2, Equations (9) and (10) of Chapter 1 must be solved. We have seen that the allowed values of the energy are given by (7), and substitution in (11) of Chapter 1 gives the ratio of the coefficients. From the values of the coefficients in the normalized MO's we can compute, for any state of the molecule, the number of electrons on each atom by multiplying (19) of Chapter 1 by the number of electrons in each MO and summing over all the MO's. Thus, q_A, the number of electrons on atom A, is given by

$$q_A = \sum_i n_i q_A^i = \sum_i n_i c_{A_i}^2 \tag{13}$$

where n_i is the number of electrons in the ith MO. Similarly, a quantity called the bond order can be defined as

$$p_{AB} = \sum n_i p'_{AB} \equiv \sum_i n_i c_{Ai} c_{Bi} \qquad (14)$$

which, as its name implies, is an indicator of the net amount of bonding between atoms A and B.

Problem 2.9. What are the charge densities and bond orders for the H_2 molecule in the ground state and in the first excited state? (In the latter state one electron is excited from the bonding to the antibonding MO.)

Problem 2.10. Consider the case of a heteronuclear diatomic with AO's in which $\alpha_1 = -12$ eV, $\alpha_2 = -4$ eV and $\beta = -3$ eV. Find the MO's and the associated energy of each. Show that the MO's are orthogonal.

Problem 2.11. Compute the charge distribution and bond order in the two lowest states of the heteronuclear diatomic of Problem 2.10. Compare with the result for H_2. Does this give you any insight into the transition from covalent to ionic bonding?

Suppose now that a diatomic is formed from two monovalent atoms, A and B, in which the difference $\alpha_A - \alpha_B$ is small compared to the magnitude of their covalent interaction. Then it is legitimate to expand the lowest root of (7) in the power series

$$E = \frac{1}{2}(\alpha_A + \alpha_B) + \beta_{AB}\left[1 + \frac{1}{2}\left(\frac{\alpha_A - \alpha_B}{2\beta_{AB}}\right)^2 + \cdots\right] \qquad (15)$$

In the approximation that total energies can be represented as sums of orbital energies, the total energy of the A—B molecule is given by $2E$. We already know that the orbital energies of the homonuclear diatomics A—A and B—B are given respectively by $E_A = \alpha_A + \beta_A$ and $E_B = \alpha_B + \beta_B$. Thus, ΔE, the change in energy for the reaction $(1/2A_2) + (1/2B_2) \rightarrow$ AB can be written as

$$\Delta E = \alpha_A + \alpha_B + 2\beta_{AB} + \frac{1}{4}\frac{(\alpha_A - \alpha_B)^2}{\beta_{AB}} - (\alpha_A + \alpha_B + \beta_A + \beta_B)$$

$$\qquad (16)$$

$$= \frac{1}{4}\frac{(\alpha_A - \alpha_B)^2}{\beta_{AB}}$$

with the reasonable assumption that $2\beta_{AB} = \beta_A + \beta_B$

Problem 2.12. Mathematically, is this assumption strictly valid if $\frac{1}{2}(S_{AA} + S_{BB}) = S_{AB}$? [Hint: Use Equation (24) of Chapter 1.]

Since the quantity $(\alpha_A - \alpha_B)^2/\beta < 0$, we would expect that the formation of a heteronuclear from two homonuclear diatomics would be exothermic, and this is almost always found to be the case. Moreover, if β is fairly constant among the heteronuclear diatomics, ΔE should be proportional to $(\alpha_A - \alpha_B)^2$. It should be possible, therefore, according to (16), to define a set of quantities x such that the difference between the bond energy of a heteronuclear diatomic A—B and the average of the bond energies of the corresponding homonuclear molecules is given by the equation

$$E(A{-}B) - \tfrac{1}{2}[E(A{-}A) + E(B{-}B)] \equiv -\Delta E = k(x_A - x_B)^2 \quad (17)$$

This is precisely the relationship that Pauling found empirically, and he gave the name electronegativities to the quantities x. Since Pauling's empirical electronegativities ought to be proportional to our α's, which in their turn are proportional to Mulliken electronegativities, we should expect a linear relationship between the empirically derived Pauling values and those defined by Mulliken from theoretical considerations. Such a relationship does, in fact, approximately hold.

Problem 2.13. Pauling attributes the exothermicity of the $(1/2A_2) + (1/2B_2) \rightarrow AB$ reactions to the contribution of the ionic resonance structure A^+B^- in the product. Explain the reason for this stabilization in the description provided by MO theory by considering qualitatively the relative electron density on the more electronegative atom in the reactant and product.

Problem 2.14. Qualitatively, why might β in a series of heteronuclear diatomics be better described as constant rather than by an equation like $\beta = k(\alpha_A + \alpha_B)$ where k is a constant? [Hint: Consider the probable relationship between the overlap integral and the value of α_A via the correlation between the relative amount of the AO ϕ_A localized close to the nucleus of atom A and the relative electronegativity of atom A.]

The Geometry of Triatomics—
Perturbation Theory

In what has become a classic series of papers, Walsh in 1953 discussed the geometry and spectra of several classes of polyatomic molecules using MO theory. We shall follow his approach in predicting the geometry of some triatomics by comparing the energies of the MO's in the linear and bent forms. Our approach will differ from that of Walsh in the following respect. Walsh followed the changes in energy of the MO's on molecular bending by considering, in addition to changes in overlaps, the changes in hybridization of the constituent AO's. We will adopt the somewhat more satisfactory and instructive procedure of using perturbation theory to examine how the energy of each of the MO's alters.

We can anticipate two types of changes in the energy of an orbital, if we treat a change in molecular geometry as a small perturbation $\mathcal{H}'$ to the Hamiltonian $\mathcal{H}$. One is due to a change in the energy $\int \psi_{i0}\mathcal{H}'\psi_{i0}\,d\tau$, of the original orbital, ψ_{i0}. The other comes from the mixing of the original orbital ψ_{i0} with another ψ_{j0} under the influence of $\mathcal{H}'$. We have already seen an example of the second kind of energy change when we considered interactions between MO's of the same symmetry in homonuclear diatomics. Here two MO's mixed under the influence of a perturbation, in this instance something we ignored when we constructed our $s\sigma_g$ and $p\sigma_g$ wavefunctions separately, namely their interaction. We could, of course, have set up the problem from the first to allow mixing of s and p orbitals. However, the treatment we chose to use had the advantage of exhibiting quite clearly the effect of the interaction, represented by β_{sp}, on the symmetry determined MO's and showing how this interaction produced MO's consisting of hybrid AO's. This is, in general, the qualitative[4] advantage of using perturbation theory; it allows us to isolate an effect and determine its consequences. The physical-organic chemist in investigating the influence of changes in substituents on the rate of a reaction is applying the philosophy of perturbation theory. However, just as a very electron withdrawing or releasing substituent can cause such a large perturbation to a reaction being studied that the mechanism changes, for the quantitative

---

[4] There is also a quantitative advantage. Since perturbation theory directly calculates energy differences, the magnitude of say a 10% error in a perturbation theory calculation will be much smaller than that which results from subtracting two large energies, each of which contains a 10% error.

application of perturbation theory to be valid, the perturbation cannot be too large.

We seek a new perturbed wavefunction of the form

$$\psi_i = \psi_{i0} + \sum_{j \neq i} c_{ij}\psi_{j0} \qquad (18)$$

in anticipation of the mixing of other zero order wavefunctions, ψ_{j0} ($j \neq i$), with ψ_{i0} under the influence of the perturbation $\mathcal{K}'$ to the zero order Hamiltonian $\mathcal{K}_0$. Substituting (18) in the Schrödinger equation with $\mathcal{K} = \mathcal{K}_0 + \mathcal{K}'$ yields,

$$\mathcal{K}\psi_i = \mathcal{K}_0\psi_{i0} + \sum_{j \neq i} c_{ij}\mathcal{K}_0\psi_{j0} + \mathcal{K}'\psi_{i0} + \sum_{j \neq i} c_{ij}\mathcal{K}'\psi_{j0}$$

$$= E_i(\psi_{i0} + \sum_{j \neq i} c_{ij}\psi_{j0}) \qquad (19)$$

Remembering that ψ_{i0} and the ψ_{j0} are wavefunctions found for $\mathcal{K}_0$ with respective energies E_{i0} and E_{j0}, (19) can be rewritten

$$E_{i0}\psi_{i0} + \sum_{j \neq i} c_{ij}E_{j0}\psi_{j0} + \mathcal{K}'\psi_{i0} + \sum_{j \neq i} c_{ij}\mathcal{K}'\psi_{j0} = E_i(\psi_{i0} + \sum_{j \neq i} c_{ij}\psi_{j0}) \quad (20)$$

Multiplying (20) through by ψ_{i0}, integrating, and using the fact that ψ_{i0} is normalized and orthogonal (See Problem 2.6) to the ψ_{j0} we get

$$E_{i0} + \int \psi_{i0}\mathcal{K}'\psi_{i0} \, d\tau + \sum_{j \neq i} c_{ij} \int \psi_{i0}\mathcal{K}'\psi_{j0} \, d\tau = E_i \qquad (21)$$

The energy E_i of the perturbed orbital ψ_i has thus been decomposed into three contributions—the unperturbed energy of the unperturbed wavefunction, the energy change in the unperturbed wavefunction due to the perturbation, and the energy change due to the change in the unperturbed wavefunction (i.e., the energy change resulting from the mixing of ψ_{i0} with the ψ_{j0}). Of course, in order to actually evaluate E_i, we need to calculate the coefficients c_{ij}. This can be accomplished by successive approximations using Equation (20), as outlined in the appendix to this chapter. As shown in the appendix, the first approximation to each of the c_{ij} is[5]

---

[5] As discussed in the appendix, better quantitative results can be obtained if, instead of the zero order energies, E_{i0} and E_{j0}, the perturbed energies correct to first order, $E_{i0} + \int \psi_{i0}\mathcal{K}'\psi_{i0} \, d\tau$ and $E_{j0} + \int \psi_{j0}\mathcal{K}'\psi_{j0} \, d\tau$ are used in the denominator of (22) and subsequent equations.

$$c_{ij} = \frac{\int \psi_{i0} \mathcal{H}' \psi_{j0} \, d\tau}{E_{i0} - E_{j0}} \tag{22}$$

The extent to which this will be a good approximation to the c_{ij} depends on the numerator's being much smaller than the denominator. This means that not only must the energy $\int \psi_{i0} \mathcal{H}' \psi_{j0} \, d\tau$ be small but also that the wavefunctions which are mixed cannot be too close in energy if perturbation theory is to be applied with the expectation of obtaining quantitatively accurate results. Using (22) in (18) we obtain the first order approximation to the perturbed wavefunctions as

$$\psi_i = \psi_{i0} + \sum_{j \neq i} \frac{\int \psi_{i0} \mathcal{H}' \psi_{j0} \, d\tau}{E_{i0} - E_{j0}} \psi_{j0} \tag{23}$$

From (21) the energy correct to second-order (the energy change in the zero-order wavefunction due to the perturbation usually is referred to as the first order correction) is

$$E_i = E_{i0} + \int \psi_{i0} \mathcal{H}' \psi_{i0} \, d\tau + \sum_{j \neq i} \frac{\left(\int \psi_{i0} \mathcal{H}' \psi_{j0} \, d\tau \right)^2}{E_{i0} - E_{j0}} \tag{24}$$

Several comments are appropriate here. We note that use of (23) and (24) requires computation of a sum of integrals. However, we need not compute all of them, since where we can make use of symmetry, we know that only those involving interaction between wavefunctions of the same symmetry will be nonzero. Only those need be calculated and summed to give the perturbed wavefunction and energy. It is also important to observe that since the numerator of the third term in (24) is necessarily positive, the sign of each term in the second-order correction to the energy of ψ_i depends on that of the denominator. If $E_{i0} > E_{j0}$, the interaction raises ψ_i in energy; however, if $E_{i0} < E_{j0}$, ψ_i is stabilized. Thus we have the rule that when two orbitals interact, the *energy of the lower decreases and that of the upper increases*, the quantum mechanical statement of the well known fact that "the rich get richer and the poor get poorer."

The way in which two wavefunctions mix also depends on the sign of $E_{i0} - E_{j0}$. If the interaction term $\int \psi_{i0} \mathcal{H}' \psi_{j0} \, d\tau$ is negative (as it will be when the mixing of wavefunctions represents an energy lowering interaction—for instance, in-phase overlap leading to increased bonding), then if $E_{i0} > E_{j0}$, ψ_{j0} is mixed into ψ_{i0} with a minus sign. If $E_{i0} < E_{j0}$, then ψ_{j0} is mixed in with a plus sign. Thus we have the second important rule of perturbation theory—*when the interaction between two orbitals is stabilizing, the orbital of higher energy is mixed into the lower with a plus sign and the lower into the upper with a minus sign.* This, of course, is why the lower orbital is stabilized and the upper destabilized; and we have already seen an example of this phenomenon in the mixing of the $s\sigma_g$ and $p\sigma_g$ orbitals. The latter, being higher in energy than the former, was mixed into it with a plus sign to give a strongly bonding MO, while the $s\sigma_g$ MO was mixed into the $p\sigma_g$ with a minus sign, which resulted in the raising of the energy of the latter.

The first-order changes in the energy of a wavefunction are somewhat easier to analyze, since they do not involve mixing of wavefunctions. The first-order changes are especially easy to dissect within the framework of Hückel theory. Let us suppose that we have already carried out a zero-order calculation and determined the coefficients in an LCAO-MO wavefunction, $\psi_{i0} = \sum_r c_{ir}\phi_r$. The energy of this wavefunction can be found from

$$E_{i0} = \int \psi_{i0}\mathcal{H}_0\psi_{i0}\, d\tau = \sum_r c_{ir} \int \phi_r\mathcal{H}_0 \sum_s c_{is}\phi_s\, d\tau \tag{25}$$

Rewriting the summation in (25), noting that $\sum\limits_{r \neq s,\, s} = 2 \sum\limits_{r > s,\, s}$

$$E_{i0} = \sum_r c_{ir}^2 \int \phi_r\mathcal{H}_0\phi_r\, d\tau + \sum_{r > s,\, s} 2c_{ir}c_{is} \int \phi_r\mathcal{H}_0\phi_s\, d\tau \tag{26}$$

Substituting for the definitions of α and β and recalling the definitions of q_r^i, the electron density at atom r, and p_{rs}^i, the bond order between atom r and atom s in the MO ψ_i, (26) becomes

$$E_{i0} = \sum_r q_r^i\alpha_r + \sum_{r > s,\, s} 2p_{rs}^i\beta_{rs} \tag{27}$$

Equation (27) gives rise to a simple expression for the first-order perturbation correction to the energy. Since in the first-order term changes in the energy due to changes in the wavefunction are neglected,[6] q_r^i and p_{rs}^i

---

[6] This does not imply that these second-order corrections are necessarily smaller than the first; in fact, we shall see that the opposite is often the case.

must remain the same. Thus the first-order change in the energy of ψ_i can be written

$$\Delta E_{i0} \equiv \int \psi_i \mathcal{K}' \psi_i = \sum_r q_r^i \Delta \alpha_r + \sum_{r>s,\,s} 2p_{rs}^i \Delta \beta_{rs} \qquad (28)$$

so that the energy change can be divided into changes in the values of α's and β's caused by the change in the Hamiltonian. In practice if one were to change an α, for instance, by substituting a different atom, one would also probably change the value of the resonance integral, β, between it and the other atoms. However, β's can be altered without affecting α's through changes in molecular geometry which alter overlaps.

Problem 2.15. Show that if the total energy is written as a sum of the orbital energies of all the electrons in a molecule that it can be expressed as

$$E_T = \sum_r q_r \alpha_r + \sum_{r>s,\,s} 2p_{rs} \beta_{rs} \qquad (29)$$

Write an expression for the first-order change in the total energy of a molecule in Hückel theory. Suppose you wanted to substitute a very electronegative atom into a molecule so as to cause the largest first-order energy change. For which atom in the molecule would you choose to substitute?

Problem 2.16. The problem of a heteronuclear diatomic A—B can also be treated through the application of perturbation theory to the homonuclear A_2 case by considering that one of the α's in a homonuclear diatomic changes by $\Delta \alpha$ in going from α_A to α_B. What is the first-order change in the energy? The second-order correction? Substitute $\alpha_B = \alpha_A + \Delta \alpha$ in (15) and compare this expression for the energy from a power series expansion with the expression from perturbation theory correct through second order. What are the perturbed wavefunctions? If B is more electronegative than A ($\Delta \alpha$ negative), how do the coefficients in the bonding MO change?

Problem 2.17. Why would you not apply perturbation theory *in the fashion suggested in Problem 2.16* to finding the MO's and energies of the heteronuclear diatomic in Problem 2.10? How could perturbation theory be applied to this problem? Use a sensible approach to finding the energy of the heteronuclear diatomic of Problem 2.10, employing second-order perturbation theory. Compare your answer with the exact one and with the one obtained by treating the molecule as suggested by Problem 2.16.

We are now in a position to use perturbation theory to examine the changes in orbital energies associated with a change in the Hamiltonian on molecular bending. Since it is only the part of the energy that changes that will interest us, it is convenient to rearrange (24) to

$$\Delta E_i = E_i - E_{i0} = \int \psi_{i0}\mathcal{3C}'\psi_{i0}\, d\tau + \sum_{j \neq i} \frac{\left(\int \psi_{i0}\mathcal{3C}'\psi_{j0}\, d\tau \right)^2}{E_{i0} - E_{j0}} \qquad (30)$$

Although (30) can be used to obtain quantitative results, here we shall employ it only in a qualitative fashion to provide insight into the factors which influence the shapes of triatomic molecules consisting of first-row elements.

First, we shall obtain the MO's for a linear triatomic. Since we need only consider the $1s$ AO's on hydrogen atoms, the easiest case to begin with is an AH_2 molecule, where A is a first row element. From symmetry $c_{i1} = \pm c_{i2}$, where the coefficients are for the two hydrogen $1s$ AO's in the ith MO. The plus combination is of σ_g symmetry and can mix with the $2s$ orbital of atom A. This mixing will, as usual, lead to two MO's, one of which, $1\sigma_g$, will be bonding between A and the two hydrogen atoms, and the other of which, $2\sigma_g$, will be antibonding. The antisymmetric combination of $1s$ hydrogen orbitals, corresponding to $c_{i1} = -c_{i2}$, is of σ_u symmetry and can mix with the $2p$ orbital of the same symmetry on A. Again the mixing will give an orbital, $1\sigma_u$, which is bonding between A and the two hydrogen atoms and another, $2\sigma_u$, which is antibonding. Finally there are two degenerate nonbonding p orbitals of π_u symmetry on A. These MO's are sketched below and a reasonable ordering of their energies is $1\sigma_g < 1\sigma_u < \pi_u < 2\sigma_g < 2\sigma_u$. The σ_g orbitals are expected to lie below the σ_u, since the former involve the $2s$ orbital on A, which is lower in energy than the σ_u $2p$. The $2\sigma_g$ and $2\sigma_u$ MO's are placed above the MO's of π_u symmetry because the 2σ orbitals are antibonding between A and the hydrogen atoms, while the π_u MO's are nonbonding.

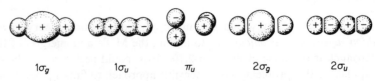

$$1\sigma_g \qquad\qquad 1\sigma_u \qquad\qquad \pi_u \qquad\qquad 2\sigma_g \qquad\qquad 2\sigma_u$$

Molecular Orbitals in a Linear AH_2 Molecule

Figure 2.4

Now we can use (30) to qualitatively predict the way the orbitals change in energy as the molecule bends. The first-order term, $\int \psi_{i0} \mathcal{H}' \psi_{i0} \, d\tau$, is easy to assess. As the molecule bends $\int \phi_1 \mathcal{H}' \phi_2 \, d\tau = \beta_{12}$ will increase in magnitude as the value of S_{12} increases. This will lower the energy of the two σ_g MO's, where the two hydrogen 1s orbitals are in phase (positive bond order), and raise the energy of the two σ_u MO's in which the interaction between these AO's is antibonding. Moreover, the energy of the bonding $1\sigma_u$ MO will be further raised on bending by the decrease in overlap between the 2p AO on A and the hydrogen 1s orbitals, caused by the directional character of the former. This decrease in overlap might be expected in the antibonding $2\sigma_u$ to offset the energy increase due to the small increase in magnitude of β_{12} and to lead to a net decrease in energy of this orbital on bending. The first order term will leave the energy of the π_u orbitals unchanged; but the p orbital which lies in the molecular plane after bending can mix with MO's that, like it, are symmetric about the remaining axis of symmetry. In other words, after bending, the bonding and antibonding interactions between this p orbital and the $1\sigma_g$ and $2\sigma_g$ MO's no longer cancel, and the resulting mixing causes changes in the energies of these orbitals via the second-order perturbation term. Although the magnitude of $\int \psi_{\sigma_g} \mathcal{H}' \psi_p \, d\tau$ may be roughly the same for the interaction of the p orbital in the plane with both $1\sigma_g$ and $2\sigma_g$, the difference in energy $E_{1\sigma_g} - E_p$ is much larger in magnitude than $E_{2\sigma_g} - E_p$. Thus, (30) shows that although mixing with $1\sigma_g$ will tend to raise slightly the energy of the p orbital, the much heavier mixing with $2\sigma_g$, because of its energetic proximity, will substantially lower the energy of the p orbital at the expense of that of $2\sigma_g$. The MO's that result from this mixing are shown schematically in Figure 2.5, and it is apparent that the energy of the p orbital is lowered because it becomes a weakly bonding hybrid.

We have now found the changes in MO energies on alteration of the molecular geometry in AH_2 by considering bending as a perturbation of the energies of the original MO's. Nevertheless, we might have proceeded in a different and, perhaps for purposes of actual computation, more gen-

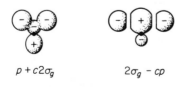

$p + c2\sigma_g$ $2\sigma_g - cp$

Figure 2.5

erally useful fashion by determining the correct MO's for the new molecular geometry and calculating their energies. However, this would not have told us how each of the old MO's transformed into each of the new ones. This information could be gleaned from construction of a correlation diagram, as explained below.

We can classify all the MO's in the product and reactant with respect to the symmetry elements that are maintained during the reaction (bending). We have already seen that the symmetry of an MO is preserved, no matter how much it mixes with other MO's of the same symmetry. Therefore, we can correlate each MO in the reactant with one of the same symmetry in the product into which it must be transformed by the reaction. If several MO's have the same symmetry, as is usually the case, the MO's are correlated in the order of their energy, lowest in the reactant with the lowest in the product, etc. We can do this because we know that two MO's of the same symmetry will always interact so that the energy of the lower is further lowered and that of the upper raised; therefore, orbitals of the same symmetry will not cross.

Problem 2.18. Construct, as we have done for the linear AH_2, the MO's for the bent molecule. Classify each orbital as S (symmetric) or A (antisymmetric) with respect to a plane containing the three atoms and to a 180° (twofold) axis of rotation which bisects the HAH angle. Do the same for the MO's in the linear molecule, and construct a correlation diagram. This diagram should, of course, be identical with Figure 2.6, derived from use of perturbation theory.

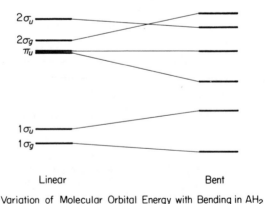

Linear   Bent

Variation of Molecular Orbital Energy with Bending in AH_2

Figure 2.6

From Figure 2.6 we can now try to make predictions regarding the geometry of AH_2 molecules in the first row. Since the directional nature of $1\sigma_u$ causes it to increase in energy on bending much faster than $1\sigma_g$ decreases, BeH_2, with four electrons, is both predicted and found to be a linear molecule. However, in its lowest excited state, where an electron is removed from $1\sigma_u$ and placed in π_u, the molecule should bend. BH_2 should be bent, and molecules with a total of 6–8 valence electrons (i.e., CH_2, NH_2 and OH_2) are all expected to be strongly bent. With one small exception these predictions are in accord with experimental fact.

Problem 2.19. A strongly bent singlet is a low-lying excited state of CH_2; the ground state is more nearly linear. Suggest an explanation.

Problem 2.20. Consider a linear triatomic AB_2 molecule, composed of first row elements. Suppose that the terminal atoms are *sp* hybridized so that each bears a σ orbital directed toward A, a nonbonding orbital directed away from A, and two *p* orbitals oriented for π bonding with A. Form combinations of these AO's proper to the molecular symmetry and give the symmetry designation of each (e.g., $p_1 + p_2$ is of π_u symmetry). What is the symmetry of each of the four AO's on A? Make a table showing how the AO's will interact to form the MO's and designate, as we did for the AH_2 case, each MO by its order in energy among MO's of its symmetry. Show that a reasonable ordering of the energy of the MO's is $1\sigma_g < 1\sigma_u < 2\sigma_g = 2\sigma_u < \pi_u < \pi_g < 2\pi_u < 3\sigma_g < 3\sigma_u$. What is the sign of the first-order change in energy of each MO on molecular bending? You may assume that the nonbonding MO's, $2\sigma_g$ and $2\sigma_u$, undergo no interactions. On bending do the MO's of π symmetry remain degenerate? Show it to be probable that when the six lowest sets of MO's are filled, there is no appreciable first order decrease in energy on molecular bending and that, if anything, a slight increase is expected. Show that the chief second-order energy change comes from the mixing of the two antibonding orbitals $2\pi_u$ and $3\sigma_g$. Through bonding interactions between which AO's do these MO's mix? Sketch the lower of the two MO's resulting from the mixing of $2\pi_u$ and $3\sigma_g$ on bending.

Problem 2.21. Predict the geometry of CO_2 and CO_2^-. How would you expect the geometry to change in the series NO_2^+, NO_2, NO_2^-? Predict the geometry of N_3^-, O_3, FO_2 and OF_2. What geometry would you expect in a trihalide ion, X_3^-, or in an "inert" gas difluoride?

Problem 2.22. In contrast to CH_2, CF_2 has a bent singlet well below any other state. Can you explain this difference? [Hint: In which molecule is the degeneracy of the π orbitals most effectively removed by bending?]

Summary

Simple MO theory, in conjunction with the application of symmetry, enabled us, in this chapter, to make accurate predictions regarding the stability and nature of the lowest electronic state of some homonuclear diatomics. In considering the mixing of wavefunctions in this problem, we were led to two important rules. (1) In a given molecular geometry wavefunctions of different symmetry do not mix. (2) When two wavefunctions do mix, the energy of the lower is further decreased while that of the upper is raised by the same amount (when overlap is neglected). We observed that the mixing of the σ MO's formed from s and p orbitals was equivalent to constructing the resulting MO's from AO's whose hybridization was optimized for bonding.

In our discussion of the geometry of triatomics we saw the importance of both the first-order term (the change in energy of the original wavefunction evaluated with the perturbed Hamiltonian) and the second-order term (the change caused by mixing of wavefunctions) in perturbation theory for estimating the net change in the energy of an MO on molecular bending. We showed that such problems could equivalently be treated by construction of a correlation diagram.

We also examined in this chapter the bonding in heteronuclear diatomics and the MO description of the transition from covalent to ionic bonding. In so doing we were able to account theoretically for the success of Pauling's empirical relationship between bond energy differences and electronegativities and for the relationship between his empirical electronegativities and those defined theoretically by Mulliken.

APPENDIX FOR CHAPTER 2

Perturbation Theory

In order to derive (22) from (20) we multiply the latter equation through by one of the ψ_{j0}, say ψ_{k0}, and integrate. Making use of orthonormality, we obtain

$$c_{ik}E_{k0} + \int \psi_{k0}\mathcal{H}'\psi_{i0}\,d\tau + \sum_{j \neq i} c_{ij} \int \psi_{k0}\mathcal{H}'\psi_{j0}\,d\tau = E_i c_{ik} \qquad (31)$$

One of the terms in the sum in (31), namely the one in which $j = k$, also contains c_{ik}. Removing it from the sum we have

$$\int \psi_{k0}\mathcal{H}'\psi_{i0}\,d\tau + c_{ik}\left(E_{k0} + \int \psi_{k0}\mathcal{H}'\psi_{k0}\,d\tau\right)$$
$$+ \sum_{j \neq k \text{ or } i} c_{ij} \int \psi_{k0}\mathcal{H}'\psi_{j0}\,d\tau = E_i c_{ik} \qquad (32)$$

which can be rearranged to

$$c_{ik} = \frac{\int \psi_{k0}\mathcal{H}'\psi_{i0}\,d\tau}{E_i - E_{k1}} + \frac{\sum_{j \neq k \text{ or } i} c_{ij} \int \psi_{k0}\mathcal{H}'\psi_{j0}\,d\tau}{E_i - E_{k1}} \qquad (33)$$

using the fact that $E_{k0} + \int \psi_{k0}\mathcal{H}'\psi_{k0}\,d\tau \equiv E_{k1}$. Two problems remain before we can solve (33) for c_{ik}. The first is that we do not know E_i. However, we can approximate it by $E_i \approx E_{i0} + \int \psi_{i0}\mathcal{H}'\psi_{i0}\,d\tau \equiv E_{i1}$, the perturbed energy correct to first-order. The justification for this approximation is that we anticipate that replacing E_i by E_{i1} in the denominator should make little difference, since the difference between them—second-

and higher order corrections to the energy—must be small compared to $E_i - E_{k1}$ if perturbation theory is to be quantitatively applicable anyway.

The second problem in solving for c_{ik} is made obvious in (33); for in order to find c_{ik}, (33) shows that we must already know all the other c_{ij}, which in their turn depend on c_{ik}. The way out of this dilemma is to solve (33) by successive approximations. As a first approximation we simply ignore the second term in (33), since we expect $c_{ij} \ll 1$. Thus we find

$$c_{ik} = \frac{\int \psi_{k0} \mathcal{H}' \psi_{i0} \, d\tau}{E_{i1} - E_{k1}} \tag{34}$$

Often it may prove to be a good approximation to substitute $E_{i0} - E_{k0}$ for $E_{i1} - E_{k1}$ in the denominator of (34). This saves us having to compute first-order corrections to the energy in problems where we expect that the second-order changes are the ones of interest. Making this substitution and noting that j and k are dummy subscripts, we see that (34) is equivalent to (22).

Problem 2.23.   How would you use (34) to get a better approximation to the true c_{ik} in (33)?

FURTHER READING

LINUS PAULING, *The Nature of the Chemical Bond*, 3rd edition, Cornell University Press, Ithaca, New York, 1960. See Chapter 3 for Pauling's discussion of electronegativity and covalency, formulated in the language of resonance theory.

Almost every elementary quantum chemistry text has a section on perturbation theory in which (24), the energy of a perturbed wavefunction correct to second-order, is derived using the Rayleigh-Schrödinger (R-S) formalism rather than the Brillouin-Wigner (B-W) approach which we used. The R-S derivation actually leads to (22), while, as we have seen the B-W treatment gives (34), which differs from (22) by the use of the energies, correct to first-order, in the denominator. For a challenge use the hint given in Problem 2.23 to obtain the next order B-W correlation and compare with R-S.

A. D. WALSH, *J. Chem. Soc.*, 2260 (1953) and following papers. This series shows how results, similar to those we found, can be obtained from the hybridization approach. Moreover, the reader may be interested in this series because contained therein are the answers to Problem 2.21. Walsh's rules continue to remain the subject of theoretical interest. For another theoretical approach see the section on vibronic mixing in Chapter 6 and the references at the end of that chapter.

3

Application of
Hückel Theory to the
Pi Systems of Unsaturated
Organic Molecules

We have developed a simple model of bonding, which, coupled with the use of symmetry and perturbation theory, enabled us to predict correctly the properties of some first row di and triatomic molecules. We now turn our attention to some larger molecules, of interest to organic chemists.

Ethylene

There is no theoretical reason why we should not carry out a complete calculation on the ethylene molecule in which we would attempt to find the symmetry-correct MO's for both the σ and π bonds. Indeed, Hoffmann has developed a computer program which makes such calculations quite easy to perform. For the most part, however, the theoretical organic chemist is interested in just the π MO's, since they are responsible for the properties characteristic of unsaturated organic molecules.

Moreover, in a planar unsaturated molecule, the π MO's will not mix with any of the σ MO's because of the difference in their symmetries. Therefore, we can rigorously separate the wavefunction for the molecule into

$$\Psi = \Psi_\sigma \Psi_\pi \qquad (1)$$

In addition, a Hamiltonian which is assumed, as we have tacitly done, to operate only on one electron at a time, contains no terms by which the wavefunction Ψ_π can affect Ψ_σ. Therefore, we can divide the Hamiltonian into

$$\mathcal{H} = \mathcal{H}_\pi + \mathcal{H}_\sigma \qquad (2)$$

where $\mathcal{H}_\sigma$ only operates on Ψ_σ and $\mathcal{H}_\pi$ on Ψ_π. Thus, it is quite legitimate to separate the problem into

$$\mathcal{H}_\pi \Psi_\pi = E_\pi \Psi_\pi \qquad (3)$$

and

$$\mathcal{H}_\sigma \Psi_\sigma = E_\sigma \Psi_\sigma \qquad (4)$$

We may anticipate that when we explicitly include the two-electron repulsion operator, e^2/r_{ij}, in the Hamiltonian, the π wavefunction will affect the σ and vice versa, via an interaction term, $\mathcal{H}_{\sigma\pi}$, in the Hamiltonian, $\mathcal{H}$. However, as we shall see in a later chapter, if we know Ψ_σ, we can treat the sigma system as merely providing part of the effective potential in which the π electrons move. The assumption that we know Ψ_σ, and that it remains the same for any state of the π system in which we are interested, is known as the π electron approximation and allows us to focus attention exclusively on the problem of solving

$$\mathcal{H}_\pi^{\text{effective}} \Psi_\pi = E_\pi \Psi_\pi \qquad (5)$$

If we make the π electron approximation for ethylene, our problem is reduced to finding the two MO's that result from interaction of the two p AO's. But we have, in fact, already solved this problem; for in Hückel theory the interaction between two carbon $2p$ AO's in a π fashion has the same description as that between two hydrogen $1s$ AO's in a σ fashion. Only the values of the parameters α and β are changed.

Problem 3.1.   Predict the result of exciting an electron from the bonding to the antibonding MO in H_2. Make the same prediction for a $\pi \longrightarrow \pi^*$ excitation in ethylene.

Allyl

In allyl the problem is to find the MO's formed from three carbon $2p\ \pi$ orbitals. The MO's will have the form

$$\psi = c_1\phi_1 + c_2\phi_2 + c_3\phi_3 \tag{6}$$

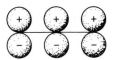

Figure 3.1

and we could find them by solving the three equations in four unknowns that result from multiplying the Schrödinger equation sequentially by ϕ_1, ϕ_2, ϕ_3 and integrating. There is, however, a much simpler way to proceed, since from symmetry $c_{i1} = \pm c_{i3}$ in any MO ψ_i. Because ϕ_2 is symmetric to the plane bisecting the $C_1 - C_2 - C_3$ angle, it will not mix with the $c_1 = -c_3$ combination which is antisymmetric. We know, therefore, from symmetry alone, that one of the MO's must be

$$\psi = \frac{1}{\sqrt{2}}(\phi_1 - \phi_3) \tag{7}$$

which has a node at the central carbon atom. Since the overlap integral S, and hence β, fall off rapidly with distance, we will approximate that β between atoms not joined by a single bond is zero. Thus, the orbital that we have just found must be nonbonding, having $E = \alpha$.

In order to find the remaining two symmetric MO's, we must find the values of c_1, c_2, and E which will satisfy

$$\mathcal{H}[c_1(\phi_1 + \phi_3) + c_2\phi_2] = E[c_1(\phi_1 + \phi_3) + c_2\phi_2] \tag{8}$$

Proceeding as usual we arrive at the equations

$$c_1\alpha + c_2\beta = c_1 E \tag{9}$$

and

$$2c_1\beta + c_2\alpha = c_2 E \tag{10}$$

Their solution gives the following MO's and the associated energies for allyl:

$$\psi_1 = \frac{1}{2}(\phi_1 + \sqrt{2}\,\phi_2 + \phi_3) \qquad E = \alpha + \sqrt{2}\,\beta$$

$$\psi_2 = \frac{1}{\sqrt{2}}(\phi_1 - \phi_3) \qquad E = \alpha$$

$$\psi_3 = \frac{1}{2}(\phi_1 - \sqrt{2}\,\phi_2 + \phi_3) \qquad E = \alpha - \sqrt{2}\,\beta$$

Let us use the results tabulated above to compare the MO picture of the allyl carbocation with that derived from resonance theory. Since two equivalent structures can be written for this carbocation, resonance theory predicts: (1) The allyl carbocation is more stable than a carbocation that is not conjugated with a double bond. (2) The positive charge is distributed equally between the two terminal atoms. (3) There is a partial double bond between the central and two terminal atoms. These predictions are virtually identical to those made by MO theory. We calculate: (1) The total energy of the two π electrons in the allyl carbocation is $2\alpha + 2\sqrt{2}\,\beta$, which is lower than the total energy, $2\alpha + 2\beta$, of the electrons in an ethylene that is not conjugated with a carbocation. (2) The π charge density at the central carbon atom is $q_2 = 2(1/\sqrt{2})^2 = 1$, while at the terminal atoms it is $q_1 = q_3 = 2(\frac{1}{2})^2 = \frac{1}{2}$; so the terminal atoms each have a net charge of $+\frac{1}{2}$. (3) The bond order between the central and terminal atoms is $p_{12} = p_{23} = 2 \times \frac{1}{2} \times 1/\sqrt{2} = 0.707$. Thus, the only difference in the conclusions drawn from the two different methods of treating the allyl carbocation is that MO theory predicts larger fractional π bonds than the half π bonds that resonance theory might lead one to expect.

Problem 3.2. Verify Equations (9) and (10). Compute the total energies, *net* charge densities (remember each atom in the σ core has a positive charge before the π electrons are added), and bond orders for the allyl radical and anion. How does the π energy of an ethylene and an isolated carbon atom compare with those above? Show that the MO picture of the stabilization and structure of these allylic systems is consistent with that derived from resonance theory.

Problem 3.3. What does first-order perturbation theory [Hint: See Equation (29) of Chapter 2] say about the change in π energy of the allyl cation and anion as the magnitude of β_{13} (remember its sign) increases? What would you predict about their relative geometries?

Problem 3.4. Consider a simplified model for a three atom transition state in which each atom has only an s valence orbital and the entering and leaving atoms are identical. Qualitatively, what are the MO's for this system? On the basis of your answer to 3.3, discuss the expected dependence of the preferred geometry on the number of electrons involved. How does the inclusion of a p orbital of σ_u symmetry on the central atom alter your conclusions? In light of the above, predict the stereochemistry of the S_E2 and S_N2 reactions (respectively two and four-electron, three-atom transition states).

Butadiene

Again symmetry can be used in order to reduce this problem of solving four equations in five unknowns to the by now familiar two equations in three unknowns. Symmetry allows us to write $c_{i1} = \pm c_{i4}$ and $c_{i2} = \pm c_{i3}$. We know that only combinations of the same symmetry will mix, so that the problem can be divided into finding MO's of the form $\psi = c_1(\phi_1 + \phi_4) + c_2(\phi_2 + \phi_3)$ which are symmetric about the plane through the C_2-C_3 bond in *cisoid* butadiene, and those of the form $\psi = c_1(\phi_1 - \phi_4) + c_2(\phi_2 - \phi_3)$, which are antisymmetric.

Figure 3.2

Problem 3.5. Try to find an MO of the form

$$\psi = c_1(\phi_1 + \phi_4) + c_2(\phi_2 - \phi_3)$$

which mixes combinations of different symmetry. Show that the equations, which are obtained for the coefficients, can be satisfied only if one coefficient is zero; (i.e., if the symmetric and antisymmetric combinations of AO's do not mix.)

The MO's and associated energies for butadiene are:

$$\psi_1 = 0.37(\phi_1 + \phi_4) + 0.60(\phi_2 + \phi_3) \qquad E = \alpha + 1.62\beta$$

$$\psi_2 = 0.60(\phi_1 - \phi_4) + 0.37(\phi_2 - \phi_3) \qquad E = \alpha + 0.62\beta$$

$$\psi_3 = 0.60(\phi_1 + \phi_4) - 0.37(\phi_2 + \phi_3) \qquad E = \alpha - 0.62\beta$$

$$\psi_4 = 0.37(\phi_1 - \phi_4) - 0.60(\phi_2 - \phi_3) \qquad E = \alpha - 1.62\beta$$

giving the bond orders for the ground state as $p_{12} = 0.89$ and $p_{23} = 0.45$. We note that there is a partial double bond between atoms 2 and 3, corresponding to a contribution to the ground state from the structure $\cdot CH_2\!-\!CH\!=\!CH\!-\!CH_2\cdot$ in resonance theory. We also observe that the total energy of butadiene is 4.48β, which is lower than that of two ethylene molecules by 0.48β. This energy lowering, or "delocalization energy," is due to the net build-up of electron density between atoms 2 and 3, as shown by the net bond order of 0.45 between these atoms.

However, if the net bond order between carbons 2 and 3 is 0.45, it is inconsistent to use the same value for β_{23} as for β_{12}, as we did in finding the MO's above. Since β depends on the overlap integral S, which depends on the bond length r, β must be a function of r. For planar pi systems the dependence of β on r is usually approximated as

$$\beta = \beta_o e^{-A(r-r_o)} \tag{11}$$

The bond length, in turn, depends on the bond order, p; in fact, the equation

$$r_{eq} = 1.52 - 0.18p \tag{12}$$

gives an excellent fit to the experimental equilibrium bond lengths between the sp^2 hybridized carbons in graphite ($p = 0.525$, $r = 1.42$ Å), benzene ($p = 0.667$, $r = 1.40$ Å), and ethylene ($p = 1.00$, $r = 1.34$ Å). (In these compounds p is completely determined by symmetry and, hence, independent of the value of β.) Our calculation on butadiene shows that the center bond should be longer than the other two, as is found experimentally; therefore, we should adjust our β's accordingly.[1] We can make our calculation self-consistent by recalculating the MO's with the new β's and repeating this procedure until the calculation converges on values of the bond lengths which are unaltered by further cycles of computation.

Problem 3.6. Suppose that in the region of interest, for small deviations of r from 1.34 Å, the bond length in ethylene, the assumed exponential dependence of β on r can be expanded in a power series as $\beta = \beta_o - 2.7(r - 1.34)\beta_0$, where β_0 is the value of the resonance integral in ethylene and the constant 2.7 Å^{-1} can be determined semiempirically, as we shall see. Use this equation and Equation (12) to calculate a new cycle of bond orders and thus a better geometry for butadiene. Note how the calculation converges by observing that, compared to the initial change from the tacitly assumed geometry in which all bond lengths, and hence all β, were equal, the geometry changes very little after another cycle of calculation. [Hint: Computations on hydrocarbons (all α equal) become much simpler if E' is defined such that $E' = E - \alpha$. Moving the energy zero in this way eliminates α from all equations.] What is the new energy of butadiene? Use Equation (29) of Chapter 2 with the new values of β to demonstrate that first order perturbation theory gives a good approximation to the new energy. This shows that, in this molecule, changes in the wavefunctions contribute very little to the change in energy with bond length. Why is it that the molecule relaxes to a geometry with a long central bond, despite the increase in π energy?

The total π energy of butadiene, once correction is made for the theoretically predicted lengthening of the central bond relative to the other two, is close to that of two ethylenes; therefore, the interaction between the two ethylene units is not expected to play an important role in the energetics of ground-state butadiene. However, electron delocalization—

---

[1] In the simplest version of Hückel theory the same value of β is used for all atoms that are nearest neighbors, regardless of the bond length between them. This poor approximation leads to quantitatively incorrect predictions (See Problem 3.6).

the fact that in some MO's substantial amounts of π-electron density is built up across the formal single bond between the two ethylene units— is predicted to play an important role in the excited-state chemistry of the molecule. First, delocalization lowers the energy difference between the highest filled and lowest unfilled MO's (1.24β, compared to 2.0β in ethylene), so that absorption of electromagnetic radiation should occur at longer wavelengths in butadiene than in ethylene. Second, in the lowest excited state the strongest bond in butadiene is between atoms 2 and 3 ($p_{23} = 0.72$) while the terminal bond orders are considerably less ($p = 0.44$). Therefore, in contrast to the ground state where rotation about the 1-2 and 3-4 bonds is difficult while about 2-3 it is relatively free, in the lowest excited state this situation should be completely reversed. The spectra and photochemistry of butadiene derivatives are generally in accord with these theoretical predictions.

Cyclobutadiene

In this molecule and in every molecule where all carbon atoms involved in bonding are equivalent, the MO's are completely determined by symmetry. From symmetry, we have the relationships $c_{i1}^2 = c_{i2}^2 = c_{i3}^2$

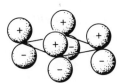

Figure 3.3

$= c_{i4}^2$. Since we now recognize that the energy of orbitals increases with the number of nodes, the MO of lowest energy in cyclobutadiene is the nodeless $\psi_1 = \frac{1}{2}(\phi_1 + \phi_2 + \phi_3 + \phi_4)$ which has energy $E = \alpha + 2\beta$. The MO of next highest energy is anticipated to have one nodal plane, but we can construct two such MO's: $\psi_2 = \frac{1}{2}(\phi_1 + \phi_2 - \phi_3 - \phi_4)$ and $\psi_3 = \frac{1}{2}(\phi_1 - \phi_2 - \phi_3 + \phi_4)$, which turn out to be degenerate in energy. Application of the integrated form of the Schrödinger equation shows that they both have $E = \alpha$ so that they are nonbonding. Moreover, they are distinct MO's; for they are orthogonal,[2] since

$$\int \psi_2 \psi_3 \, d\tau = \tfrac{1}{4} \int (\phi_1^2 - \phi_2^2 + \phi_3^2 - \phi_4^2) \, d\tau = 0.$$

As was the case with the degenerate pair of orthogonal atomic orbitals of π_u symmetry in a linear AH_2 molecule, these MO's in cyclobutadiene do not interact.

$$\int \psi_2 \mathcal{H} \psi_3 \, d\tau = \tfrac{1}{4} \int (\phi_1 + \phi_2 - \phi_3 - \phi_4) \mathcal{H} (\phi_1 - \phi_2 - \phi_3 + \phi_4) \, d\tau$$

$$= \tfrac{1}{4}[2(\alpha - \alpha) + 4(\beta - \beta)] = 0 \tag{13}$$

Therefore, since these MO's have the same energy and do not interact we can construct two new MO's of the same energy from the orthogonal combinations

$$\psi_2' = \psi_2 \sin \theta + \psi_3 \cos \theta \quad \text{and} \quad \psi_3' = \psi_2 \cos \theta - \psi_3 \sin \theta$$

These new MO's are also normalized since

$$\int \psi_2' \psi_2' \, d\tau = \sin^2 \theta \int \psi_2 \psi_2 \, d\tau + \cos^2 \theta \int \psi_3 \psi_3 \, d\tau$$

$$= (\sin^2 \theta + \cos^2 \theta) = \int \psi_3' \psi_3' \, d\tau = 1 \tag{14}$$

Because θ can assume an infinite number of values, there are an infinite number of sets of degenerate MO's.

Problem 3.7. Show that $\psi = \tfrac{1}{2}(\phi_1 + \phi_2 + \phi_3 - \phi_4)$ is not an acceptable MO (a) from the point of view that orbitals are standing waves of defined wavelength, (b) from the point of view that orbitals should have a defined symmetry with respect to the elements of symmetry present in the molecule, and (c) from the point of view that ψ does not satisfy the Schrö-

---

[2] The reader may care to prove for himself that since the wavefunctions are respectively SA and AS with respect to the two mirror planes, m_1 and m_2, through the bonds, they remain orthogonal even when the approximation that $S = 0$ is not made.

dinger equation, $\mathcal{H}\psi = E\psi$. [Hint: Show that multiplication by different AO's followed by integration gives different values of E.] Show how to find the correct wavefunction for the highest MO.

Shown in Figure 3.4 are the MO's (one of the infinite number of equivalent ψ_2, ψ_3 pairs is shown) for cyclobutadiene, into which we must

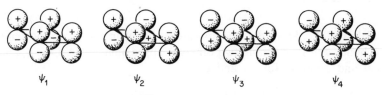

$$\psi_1 \qquad \psi_2 \qquad \psi_3 \qquad \psi_4$$

π MO's For Cyclobutadiene

Figure 3.4

feed four electrons. Two will go into ψ_1 and we are left with two to put into the degenerate ψ_2 and ψ_3. If we were to follow Hund's rule, which we should remember was derived for atoms and so need not necessarily apply to molecules, we would place the electrons with parallel spins, one in each MO. Thus, cyclobutadiene would be a ground state triplet molecule—a rather strange result for a stable hydrocarbon. The total energy also leads us to question the stability of cyclobutadiene, for it is $4\alpha + 4\beta$, the same as that of two isolated ethylenes. But the energetic situation is even worse than that. The bond orders are only 0.5; therefore, from (12) the bond length should be 1.43 Å, and from (11), using $A = 2.7$ Å^{-1}, the value of β is computed to be only $0.78\beta_o$, where β_o is the value of β for ethylene. Thus, cyclobutadiene is actually expected to be some $-0.88\beta_o$ less stable than two ethylene molecules; and despite innumerable attempts to prepare cyclobutadiene, the molecule has only recently been isolated at very low temperatures in an inert matrix.[3,4]

Of course, cyclobutadiene could shorten two bonds and lengthen two others in order to achieve the stability of two ethylenes; but this route

---

[3] This is due in part to the ability of the molecule to undergo facile reactions, a property, as we shall see in a later chapter, also conferred on it by the electronic nature of its π system.

[4] For leading references to recent (at the time of this writing) developments in cyclobutadiene chemistry see *J. Amer. Chem. Soc.*, **95**, 8481 (1973).

is not open to the triplet, for the decrease in energy of one degenerate orbital on such a molecular distortion is matched by an increase in the energy of the other. However, for a singlet state where both electrons occupy either ψ_2 or ψ_3, the molecule might be expected to distort from a square to a rectangle. Such a distortion, which lifts the degeneracy of an electronic state by lowering the energy of one of the two degenerate orbitals, is called the Jahn-Teller effect after the physicists responsible for its theoretical description.

Problem 3.8.   Show that the first-order change in energy on shortening two bonds and lengthening two others is approximately zero for ψ_1, and equal and opposite for ψ_2 and ψ_3.

Thus, we would predict that cyclobutadiene will be square if it is a triplet and rectangular if it is a singlet. Since the stablization of the triplet comes from correlating the movement of electrons, about whose repulsive interaction our Hamiltonian is as yet uninformed, we have no way at the moment of predicting whether a square triplet or a distorted singlet is more likely to be the actual ground state. In a later chapter when we explicitly include electron repulsion in the Hamiltonian, we shall be able to make this prediction. Moreover, we will see that although the two orbitals ψ_2 and ψ_3 are degenerate, the lowest singlet state is not. Distortion can nevertheless occur through mixing with another low-lying singlet to give a "pseudo Jahn-Teller effect."

Since we cannot now attempt the quantitative resolution of the problem of the relative stability of the singlet and triplet states, let us examine qualitatively how we might go about stabilizing the former with respect to the latter in an undistorted cyclobutadiene derivative. Hoffmann has shown how this might be accomplished, and his solution provides an excellent exercise in the qualitative use of perturbation theory and symmetry arguments. Suppose we introduce four substituents onto the ring, two of which are π electron donating, possessing a filled π orbital lying below ψ_2 and ψ_3, the other two being electron withdrawing by virtue of an empty π orbital lying above the degenerate cyclobutadiene pair. Following Hoffmann, let us call the latter substituents X and the former Y. The problem now is to decide how to arrange these substituents on the ring in order to effect maximum lifting of the degeneracy. There are two possibilities:

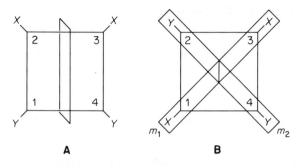

A   **B**

Figure 3.5

In **A** we can classify the π orbitals of the X and Y substituents according to their symmetry with respect to the plane that bisects the molecule vertically. The in-phase combination of the π_x orbitals is S, while the out-of-phase combination is A; the same holds for the π_y. Since neither the X's nor the Y's are bonded together, both in-phase and out-of-phase combinations will have approximately the same energy. Therefore, the interaction of the S combination of π_y with ψ_3, also of S symmetry, will raise the energy of ψ_3 by the same amount that the mixing of the A combination with ψ_2 will raise that of ψ_2. The energy lowering of ψ_2 and that of ψ_3 through interaction respectively with the A and S combinations of π_x is also expected to be equal. Therefore, no lifting of the degeneracy of ψ_2 and ψ_3 is expected from the substitution pattern **A**.[5]

In **B** we use the two planes of symmetry to classify the orbitals. Both the in- and out-of-phase combinations of the π_x orbitals are S with respect to plane 1; thus, we get an SS and an SA combination. For the π_y orbitals we get SS and AS. The particular choice we have made for ψ_2 and ψ_3 is not a good one for this symmetry. The two diagonal planes of symmetry transform ψ_2 and ψ_3 into each other.[6] Therefore, we cannot classify these two orbitals as S or A with respect to the diagonal planes. However, we are free to make new, completely equivalent, linear combinations which

---

[5] If we were to carry our perturbation treatment to higher order to include the mixing of ψ_1 with ψ_3 and ψ_2 with ψ_4 that takes place when cyclobutadiene is substituted as in A, we would find that the new ψ_2 and ψ_3 MO's that result are no longer degenerate. Can you predict which will be lower in energy?

[6] This is the chief reason why a complete symmetry description of degenerate orbitals must be applied to the orbitals collectively instead of individually.

we can so classify. We can write

$$\psi'_2 = \frac{1}{\sqrt{2}}(\psi_2 + \psi_3) = \frac{1}{\sqrt{2}}(\phi_1 - \phi_3)$$

and

$$\psi'_3 = \frac{1}{\sqrt{2}}(\psi_2 - \psi_3) = \frac{1}{\sqrt{2}}(\phi_2 - \phi_4)$$

The former is SA and the latter AS. Thus, the energy of ψ'_2 will be lowered by its mixing with the π_x orbital of SA symmetry, while the energy of ψ'_3 will be raised by its interaction with the AS combination of π_y orbitals; no other interactions will occur.

Problem 3.9.   Which two AO's are interchanged by reflection through diagonal plane 1? diagonal 2? Show that the operation, m_1, of reflection through plane 1 gives the results $m_1\psi_2 = \psi_3$ and $m_1\psi_3 = \psi_2$. Show that $m_2\psi_2 = -\psi_3$ and $m_2\psi_3 = -\psi_2$. Demonstrate, therefore, that

$$m_1\psi'_2 = m_1 \frac{1}{\sqrt{2}}(\psi_2 + \psi_3) = \frac{1}{\sqrt{2}}(\psi_3 + \psi_2) = \psi'_2$$

and that

$$m_1\psi'_3 = m_1 \frac{1}{\sqrt{2}}(\psi_2 - \psi_3) = \frac{1}{\sqrt{2}}(\psi_3 - \psi_2) = -\psi'_3$$

What are the results for the corresponding m_2 operations?

Figure 3.6 summarizes our findings for substitution patterns **A** and **B**. In **B** the degeneracy of the two nonbonding cyclobutadiene MO's is lifted and two electrons will be placed in ψ'_2. Moreover, in **B** two empty MO's interact maximally with two filled MO's of lower energy, thus lowering the total energy and stabilizing the molecule more than in **A**. Recently Gompper and Seybold reported the synthesis and isolation of a type **B** cyclobutadiene with $X = CO_2CH_2CH_3$ and $Y = N(CH_2CH_3)_2$.[7]

---

[7] The lifting of the degeneracy does not guarantee that the substituted molecule will be more stable as a square than as a rectangle, however. In fact, the molecule prepared by Gompper and Seybold appears to undergo rapid equilibration between its two equivalent rectangular forms through a square geometry that is slightly higher in energy [*J. Amer. Chem. Soc.* **95**, 8479 (1973)].

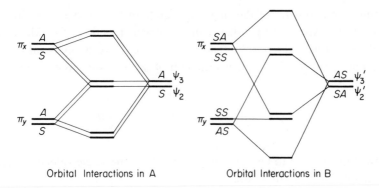

Orbital Interactions in A   Orbital Interactions in B

Figure 3.6

Problem 3.10. Suppose you wanted to obtain a stable heterocyclic cyclobutadiene containing two boron ($\alpha_B > \alpha_C$) and two nitrogen atoms ($\alpha_N < \alpha_C$). Use first-order perturbation theory to determine which of the two possible isomers you would attempt to synthesize. Determine the effect of orbital mixing on the less stable isomer.

Benzene

We have found that despite the fact that two equivalent resonance structures can be drawn for a square cyclobutadiene, the π energy of the molecule is predicted to be greater than that of two isolated ethylenes, a prediction which is in agreement with the experimental evidence of the great instability of this molecule. This correct prediction was a great success for Hückel MO theory; for resonance theory predicts that cyclobutadiene and benzene should enjoy similar resonance stabilization, since two equivalent resonance structures can be drawn for each. We shall now see that HMO theory also predicts the experimentally found stability of benzene relative to three noninteracting ethylenes.

All the carbons in benzene are equivalent, and the MO's are again completely determined by symmetry. We might once more use the fact that equivalent atoms must bear equivalent charges in each MO; but as we shall see in a later chapter, in rings other than those of $4n$ atoms this approach leads to coefficients which are complex numbers. The complex coefficients only turn up in degenerate orbitals, so we can always find real orbitals by taking appropriate linear combinations. However, let us try to find the real orbitals directly by treating the MO's as standing waves around the perimeter of the ring. The equation for a wave on a circular ring can be expressed by either a sine or cosine function of θ, where the angle θ is measured from the center of the ring between any point on the ring and atom N. If λ is the wavelength of ψ and c the circumference of the ring,

$$\psi = \sin \frac{c}{\lambda}\theta \quad \text{and} \quad \psi' = \cos \frac{c}{\lambda}\theta \tag{15}$$

Now if the wave is to be a standing one, it must be single valued at every point on the ring; therefore

$$\sin \frac{c}{\lambda}(\theta + 2\pi) = \sin \frac{c}{\lambda}\theta \tag{16}$$

This is only satisfied if

$$2\pi \frac{c}{\lambda} = 2m\pi \qquad (m = 0, 1, 2, \ldots) \tag{17}$$

so that

$$\frac{c}{\lambda} = m \qquad (m = 0, 1, 2, \ldots) \tag{18}$$

This is just the condition that an integral number of wavelengths must fit on the ring. We now wish to find the relative amplitude of ψ_m at atom r and thus determine the relative value of the coefficient c_{mr} in the LCAO wavefunction. The position of atom r corresponds to a value of $2\pi r/N$ for θ, when there are N atoms in the ring. Substituting in (16) and equating the amplitude of ψ_m at r with c_{mr}, we finally arrive at the expressions for the coefficients in the sine and cosine MO wavefunctions.

$$c_{mr} = \sin \frac{2\pi r}{N}m \quad \text{and} \quad c_{m'r} = \cos \frac{2\pi r}{N}m' \tag{19}$$

Problem 3.11.   Show that these equations give the unnormalized MO's of cyclobutadiene.

The energy of each MO, $\psi_m = \sum_r c_{mr}\phi_r$, in a cyclic polyene can be conveniently found from (19). For instance, using the MO's with coefficients given by the sine solution, applying the Schrödinger equation, multiplying by ϕ_r, and integrating, gives

$$\alpha \sin \frac{2\pi rm}{N} + \beta\left(\sin \frac{2\pi(r + 1)m}{N} + \sin \frac{2\pi(r - 1)m}{N}\right) = E \sin \frac{2\pi rm}{N}$$

(20)

since ϕ_r will interact only with the AO's ϕ_{r+1} and ϕ_{r-1} on the two atoms bonded to r. Division by $\sin (2\pi rm/N)$ yields

$$E = \alpha + \beta\left(\frac{\sin (r + 1)\omega + \sin (r - 1)\omega}{\sin r\omega}\right)$$

(21)

where $\omega = 2\pi m/N$. Using the trigonometric sums to products formulas, (21) can be rewritten

$$E = \alpha + \beta\left(\frac{\sin r\omega \cos \omega + \cos r\omega \sin \omega + \sin r\omega \cos \omega - \cos r\omega \sin \omega}{\sin r\omega}\right)$$

(22)

which reduces to

$$E = \alpha + 2\beta \cos \frac{2\pi m}{N}$$

(23)

for orbitals where $m \neq 0$ or $\frac{1}{2}N$. The same expression is obtained for the energy of the MO's given by the cosine solution, but here no restriction is placed on m. All the allowed energy levels, some of which are degenerate, are generated by $m = 0, 1, 2, \ldots N - 1$ in (23). Since for $\cos \omega$ the intervals $(0, 2\pi)$ and $(-\pi, +\pi)$ are equivalent, when N is even, the energy levels can also be generated by $m = 0, \pm1, \pm2, \pm \ldots \pm N/2$. The orbitals and their energies in benzene, $N = 6$, are given in the following table.

$$\psi_6 = \frac{1}{\sqrt{6}}(\phi_1 - \phi_2 + \phi_3 - \phi_4 + \phi_5 - \phi_6) \qquad E = \alpha - 2\beta$$

$$\psi_5 = \frac{1}{2\sqrt{3}}(2\phi_1 - \phi_2 - \phi_3 + 2\phi_4 - \phi_5 - \phi_6) \qquad E = \alpha - \beta$$

$$\psi_4 = \frac{1}{2}(\phi_2 - \phi_3 + \phi_5 - \phi_6) \qquad\qquad E = \alpha - \beta$$

$$\psi_3 = \frac{1}{2\sqrt{3}}(2\phi_1 + \phi_2 - \phi_3 - 2\phi_4 - \phi_5 + \phi_6) \qquad E = \alpha + \beta$$

$$\psi_2 = \frac{1}{2}(\phi_2 + \phi_3 - \phi_5 - \phi_6) \qquad\qquad E = \alpha + \beta$$

$$\psi_1 = \frac{1}{\sqrt{6}}(\phi_1 + \phi_2 + \phi_3 + \phi_4 + \phi_5 + \phi_6) \qquad E = \alpha + 2\beta$$

We note that two electrons can be placed in the lowest MO and the remaining four in the lowest degenerate pair, giving a closed shell. In addition, the total π energy of this electronic configuration is 8β, 2β lower than the π energy of three ethylenes. We are thus able to predict some special stability for benzene.

Problem 3.12.   Correct the delocalization energy of benzene for the fact that β in benzene is not equal to β_o for ethylene.

"Aromaticity" and the 4n + 2 Rule

We have investigated two different cyclic polyenes. We found that cyclobutadiene, if it is a singlet, should distort to a rectangular geometry because of its half-filled shell. In the most symmetrical geometry, the π energy is greater than that of the two isolated ethylenes into which the system may be formally dissected. In contrast, benzene is predicted to have a symmetrical structure, because of its filled shell; and it possesses a large stabilization energy relative to three isolated ethylenes. Cyclobutadiene has the properties that we shall associate with an "antiaromatic" system, while benzene possesses those we will identify with an "aromatic" compound.

From either the nodal properties of the MO's derived from (19), or from Equation (23) we can see that in a monocyclic polyene the MO's occur in degenerate pairs except for the lowest and, in rings comprised of even numbers of atoms, the highest. In fact, Equation (23) can be interpreted as giving the points of intersection between a circle of radius 2β whose center is at α and a regular polygon inscribed therein with a vertex

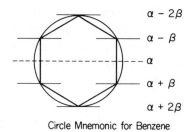

$$\alpha - 2\beta$$
$$\alpha - \beta$$
$$\alpha$$
$$\alpha + \beta$$
$$\alpha + 2\beta$$

Circle Mnemonic for Benzene

Figure 3.7

at 2β. As an example of the use of this circle mnemonic, the energy levels in benzene are pictured in Figure 3.7. Because in a monocyclic polyene, after the lowest MO has been filled, each succeeding shell of degenerate MO's can accommodate four electrons, for a filled shell a total of $4n + 2$ electrons is required.

We must defer until later in this chapter a proof that $4n + 2$ electron systems also satisfy our second criterion for aromaticity, namely, a decrease in π energy on electron delocalization. Nevertheless, it is instructive to examine the specific example of cyclopentadienyl, since in this molecular framework, we would expect the anion to be aromatic, because it has six π electrons, while the cation, with four, should be antiaromatic. In this case we must be careful to choose a realistic reference compound. For instance, in comparing the energy of benzene to that of three isolated ethylenes, we know that the delocalization, or as it is sometimes called, "resonance" energy we so compute is largely a property of the cyclic array and not just the linking of the ethylenes. We can conclude this because we found that when ethylenes are joined linearly, as in butadiene, after we have made our calculation self-consistent, the delocalization energy is small. It would, however, be misleading to choose for the calculation on cyclopentadienyl two ethylenes and an isolated carbon atom as reference; for we already have seen that, unlike the coupling of two ethylenes to form butadiene, the joining of a carbon atom to an ethylene to give an allylic system has a large delocalization energy associated with it. Therefore, if we are to avoid including this delocalization energy in the one we calculate for the cyclic system, we had better choose as our reference allyl plus ethylene, or even better, acyclic pentadienyl.

The energy of pentadienyl, assuming all β equal, can be calculated and is found to be $4\alpha + 5.46\beta$ for the cation and $6\alpha + 5.46\beta$ for the anion, since the third MO in this system is nonbonding. We wish to put these energies on a β_o scale so that we can compare them to the corresponding cyclopentadienyl energies on the same scale; for in all these molecules the bond lengths are expected to be longer than in ethylene, and the magnitude of β should be adjusted accordingly. We can combine (12) with the expanded form of (11), used in Problem 3.6, to give

$$\beta = \beta_o - 0.48(1 - p)\beta_o \qquad (24)$$

If we now knew the AO coefficients in the MO's, we could use them to compute the bond orders for the two different types of bonds in pentadienyl. Use of (24) would then allow us to calculate the two required β's, and application of first order perturbation theory would allow us to recalculate the total energy.

However, without knowing the actual MO's we can proceed in an instructive way with the information already in hand to obtain a good approximation to the π energy on the β_0 scale. From Problem 2.15, we know that the total π energy of a molecule can be written

$$E = \sum_r q_r\alpha_r + 2\sum_{r>s,\,s} p_{rs}\beta_{rs} \qquad (25)$$

We also know the total energy of the two pentadienyl ions when all β are assumed equal. Therefore, we can use (25) to compute an average bond order $\bar{p} = \frac{1}{2}(p_1 + p_2) = 0.68$ for pentadienyl. We use this in (24) to compute an average $\bar{\beta} = \frac{1}{2}(\beta_1 + \beta_2) = 0.85\beta_o$. What we actually need to compute the true energy is the quantity $4(p_1\beta_1 + p_2\beta_2)$, but all that we can calculate now is

$$8\bar{p}\bar{\beta} = 2(p_1\beta_1 + p_2\beta_2 + p_1\beta_2 + p_2\beta_1) \qquad (26)$$

Let us compute how large an error we will make if we use (26) instead of the correct expression. If we write the difference $p_1 - p_2 = \Delta$, then from (24) $\beta_1 - \beta_2 = 0.48\Delta\beta_o$. Substituting in (26),

$$8\bar{p}\bar{\beta} = 2[p_1\beta_1 + p_2\beta_2 + (p_2 + \Delta)\beta_2 + (p_1 - \Delta)\beta_1]$$

$$= 4(p_1\beta_1 + p_2\beta_2) + 2\Delta(\beta_2 - \beta_1) \qquad (27)$$

$$= 4(p_1\beta_1 + p_2\beta_2) - 0.96\Delta^2\beta_o$$

This differs from $4(p_1\beta_1 + p_2\beta_2)$ by $-0.96\Delta^2\beta_o$; and even if the two bond orders in pentadienyl differ as much as $\Delta = 0.3$, it will make our calculated energy no more than $-0.1\beta_o$ too high. Therefore, the energies of the pentadienyl cation and anion are to within $-0.1\beta_o$, respectively $E^+ = 4\alpha + 4.64\beta_o$ and $E^- = 6\alpha + 4.64\beta_o$.

Problem 3.13. Compute the energies of the allyl cation and anion on the β_o scale. What do you conclude about the stabilization of an allylic system by an additional vinyl group compared to the stabilization of a carbon atom by one?

For cyclopentadienyl we can obtain the orbital energies from (23), and we calculate the total energy of the cation to be $E^+ = 4\alpha + 5.24\beta$ while that of the anion is $E^- = 6\alpha + 6.48\beta$. Using (25), which we can now do rigorously since all the bonds are equivalent, the bond orders are 0.524 and 0.648 respectively. From these values computation of the total energies on the β_o scale gives $E^+ = 4\alpha + 4.04\beta_o$ and $E^- = 6\alpha + 5.38\beta_o$. We see, therefore, that the $4n + 2$ rule correctly predicts that the cyclopentadienyl cation is less stable than the acyclic system, while the anion is considerably more stable.

A little analysis shows why this is the case. In the acyclic cation the four electrons exactly fill the bonding MO's, making maximum use of the bonding potential of this array of atoms. This also gives maximum bond orders, hence short bonds and a high magnitude of the average β. By contrast, in the cyclopentadienyl cation a bonding MO at the same energy as a filled orbital is left empty. Thus maximum advantage is not taken of the bonding potential, and the small bond order results in a β of small magnitude.[8] In the anions the situation is reversed. Full advantage is taken of the cyclic system's capability to accept two more electrons in an orbital of the same energy as the uppermost filled in the cation, while in the acyclic system there is no bonding orbital available for these electrons. In addition, the cyclic system is further stabilized by the increase in bond orders and the consequent increase in the magnitude of β due to the shortened bond lengths. Thus, the unusual stability of monocyclic polyenes

---

[8] Both the cyclic and acyclic cation have similar total bond orders as can be inferred, by the use of (25), from their very similar uncorrected total energies of 5.24 and 5.46β, respectively. However, each of the bonds in the former has a lower bond order than the average in the latter, because the total bond order is distributed over one more bond. This results in a smaller value of β in the cyclic than in the acyclic cation, which is responsible for the greater part of the 0.60β_o difference in their total energies.

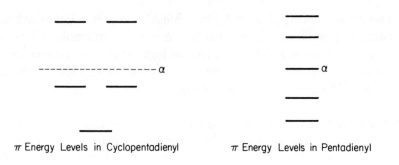

Figure 3.8

containing $4n + 2$ electrons results from the maximum advantage taken of the special bonding potentialities of the cyclic array, while in $4n$ electron systems the bonding capability of the corresponding acyclic reference system is better exploited.

Effect of the Sigma Bonds on the Total Energy[9]

There is a factor we have not yet explicitly considered in our calculations. We have been making the tacit assumption that the energy of the σ system remains unchanged; yet we have, via the use of the bond order–bond length correlation (12), been allowing the molecule to adjust its geometry. In the case of butadiene we saw that such an adjustment is made not because it lowers the π energy. On the contrary, the π energy increased when we allowed the molecule to lengthen its bonds, especially the central one, from the 1.34 Å ethylene value at which we started the calculation. Therefore, the bond order–bond length correlation must work because the energy of the σ system is lowered on bond lengthening. This is not surprising, since according to (12), the natural single bond length between sp^2 carbons is 1.52 Å. Compressing σ single bonds to the shorter lengths found in compounds with partial double bonds must increase the σ energy. However, this compression allows better π overlap, thus decreasing the π energy. A compromise between the σ and π energies is reached which makes the total energy a minimum at the equilibrium bond distance.

In our calculations of resonance energies we have not included the contribution of the changes in σ energy. Therefore, if we were to actually

---

[9] A mastery of the mathematics contained in this section is not essential for the understanding of subsequent chapters.

compare the experimental energy difference between two compounds with that which we have calculated, we would expect to find agreement only if the net σ compression energies in both were the same through a series of fortuitous cancellations. We should, therefore, correct our calculated resonance energies for sigma bond compression. Although this could be done by making use of empirical force constants in some assumed potential function, it is both instructive and also more in keeping with our almost wholly theoretical approach actually to derive an expression for the change in σ energy with bond length. From the preceding discussion we can write the condition for static equilibrium in a bond as

$$\frac{\delta E}{\delta r} = 0 = \frac{\delta E_\pi}{\delta r} + \frac{\delta E_\sigma}{\delta r} \qquad (r = r_{eq}) \tag{28}$$

With the assumption that for a small change in bond length in a *given* molecule, the wavefunction and, hence, the bond order change very little, first order perturbation theory can be applied to (25) in order to obtain an expression for $\delta E_\pi/\delta r$. Equation (28) can then be rewritten

$$2p\frac{\delta \beta}{\delta r} + \frac{\delta E_\sigma}{\delta r} = 0 \qquad (r = r_{eq}) \tag{29}$$

Problem 3.14.   Show that if β has the form suggested in (11),

$$\beta = \beta_o e^{-A(r-r_o)}$$

where A is a constant to be determined, β_0 the ethylene resonance integral, and $r_0 = 1.34$ Å, and if the sigma energy can be written as a Hooke's law potential,

$$E_\sigma = \frac{1}{2}k_\sigma(r'_0 - r)^2 \tag{30}$$

where $r'_0 = 1.52$ Å, the uncompressed natural single bond length between sp^2 carbons, then from (29) it can be derived that

$$r = r'_0 + \frac{2A\beta}{k_\sigma}p \tag{31}$$

This equation has the same form as the empirical relationship (12) between bond length and bond order, provided, of course, that over the small range (0.08 Å) where (12) is known to be valid, the nonlinearity of (31), due to the dependence of β on r, passes undetected. Use (12) to find k_σ.

We need not make assumptions like the one in Problem 3.14, about the form of the sigma energy function, since we can obtain it directly from (29). Equation (29) can be rearranged to

$$p = \frac{-\frac{1}{2}\left(\frac{\delta E_\sigma}{\delta r}\right)}{\frac{\delta\beta}{\delta r}} \qquad (r = r_{eq}) \tag{32}$$

Since the right-hand side of (32) is a function of r_{eq}, this equation implies that there exists a functional relationship between the *equilibrium* bond length of a molecule and the bond order. Thus, (32) predicts what the empirical (12) confirms—for a *series of bonds* the bond order and the equilibrium bond length are related. Therefore, in (29), which is valid for and only for a bond at its equilibrium bond length, we can substitute for p the function $P(r_{eq})$, the empirical bond order—bond length relationship.

$$\left(\frac{\delta E_\sigma}{\delta r}\right)_{r_{eq}} = -2P(r_{eq})\left(\frac{\delta\beta}{\delta r}\right)_{r_{eq}} \tag{33}$$

Making this substitution from (12) and assuming β has the form of (11), Equation (29) can be rewritten

$$\left(\frac{\delta E_\sigma}{\delta r}\right)_{r_{eq}} = +\frac{2}{0.18}(1.52 - r_{eq})A\beta_o e^{-A(r_{eq}-1.34)} \tag{34}$$

E_σ can be found by integrating (34) by parts. If 1.52 Å is taken as the bond length at which the sigma energy is zero, the reader may check by differentiation that the solution is

$$(E_\sigma)_{r_{eq}} = -11.1\beta_o(1.52 - r_{eq})e^{-A(r_{eq}-1.34)}$$
$$+ \frac{11.1}{A}\beta_o[e^{-A(r_{eq}-1.34)} - e^{-A(1.52-1.34)}] \tag{35}$$

It is interesting to examine the behavior of (35) around $r_{eq} = 1.34$ Å.

$$E_\sigma = -11.1\beta_0(1.52 - 1.34) + \frac{11.1}{A}\beta_0[1 - e^{-A(1.52-1.34)}] \tag{36}$$

Expanding the exponential in a power series,[10] Equation (36) may be written

$$E_\sigma = -11.1\beta_0(1.52 - 1.34) + \frac{11.1}{A}\beta_0[1 - 1 + A(1.52 - 1.34)$$

$$-\frac{1}{2}A^2(1.52 - 1.34)^2 + \cdots] \tag{37}$$

$$= \frac{-11.1}{A}\beta_0\left[\frac{1}{2}A^2(1.52 - 1.34)^2 + \cdots\right]$$

and we note that at least at 1.34 Å (37) does have approximately the Hooke's law form of (30), with k_σ as suggested by comparison of (31) and (12).

Provided we have a value for A, we can now make the necessary corrections to our calculated resonance energies, caused by the difference between E_σ for the most symmetrical molecular geometry and E_σ for the geometry in which single and double bonds alternate. Up to now, we have used a value of 2.7 Å^{-1} for A; we will now show that this is a reasonable value by calculating a quantity in terms of A which can be compared with an experimental figure. Thus, we may be said to be determining A "semi-empirically," since we are neither calculating it *ab initio* nor measuring it directly.

The second derivative of the bond energy, evaluated at r_{eq}, is equal to the vibrational force constant, k, for a bond, which can be found from analysis of the infrared and Raman spectra of a molecule.

$$k = \left(\frac{\delta^2 E}{\delta r^2}\right)_{r_{eq}} = 2P(r_{eq})\left(\frac{\delta^2\beta}{\delta r^2}\right)_{r_{eq}} + \left(\frac{\delta^2 E_\sigma}{\delta r^2}\right)_{r_{eq}} \tag{38}$$

The second term in (38) may be obtained from (33). Since $\delta E_\sigma/\delta r$ is not a function of r_{eq}, we can write

$$\left(\frac{\delta^2 E_\sigma}{\delta r^2}\right)_{r_{eq}} = \frac{\delta}{\delta r_{eq}}\left(\frac{\delta E_\sigma}{\delta r}\right)_{r_{eq}}$$

$$= -2\frac{\delta P(r_{eq})}{\delta r_{eq}}\left(\frac{\delta\beta}{\delta r}\right)_{r_{eq}} - 2P(r_{eq})\frac{\delta}{\delta r_{eq}}\left(\frac{\delta\beta}{\delta r}\right)_{r_{eq}} \tag{39}$$

[10] The formula for the power series expansion of e^{-x}, $x < 1$, is

$$e^{-x} = 1 - x + \tfrac{1}{2}x^2 + \ldots$$

Using the fact that $\delta\beta/\delta r$ also is not a function of r_{eq}

$$\frac{\delta}{\delta r_{eq}}\left(\frac{\delta\beta}{\delta r}\right)_{r_{eq}} = \left(\frac{\delta^2\beta}{\delta r^2}\right)_{r_{eq}} \tag{40}$$

as the reader may easily verify for himself, using (11). Equation (39) can now be rewritten

$$\left(\frac{\delta^2 E_\sigma}{\delta r^2}\right)_{r_{eq}} = -2\frac{\delta P(r_{eq})}{\delta r_{eq}}\left(\frac{\delta\beta}{\delta r}\right)_{r_{eq}} - 2P(r_{eq})\left(\frac{\delta^2\beta}{\delta r^2}\right)_{r_{eq}} \tag{41}$$

Substituting this equation into (38) gives the theoretical expression[11] for the experimental vibrational force constant, k, for a bond

$$k = -2\frac{\delta P(r_{eq})}{\delta r_{eq}}\left(\frac{\delta\beta}{\delta r}\right)_{r_{eq}} \tag{42}$$

Using (11) and (12) to obtain the required derivatives yields

$$k = \frac{-2A}{0.18}\beta_o e^{-A(r_{eq}-1.34)} \tag{43}$$

The value of k in the symmetric bond stretching mode in benzene is 7.62 $\times$ 10^5 dynes/cm. The corresponding value for the antisymmetric vibrational mode, where one set of bonds lengthens while another shortens, is 3.94 $\times$ 10^5 dynes/cm. The theoretical expression for the force constant, k', in this latter mode can be derived in a fashion analogous to (43)[12] and is

$$k' = -2A^2\left(\frac{1}{0.18A} - 1\right)\beta_o e^{-A(r_{eq}-1.34)} \tag{44}$$

---

[11] In the event that the distinction between p, which we assume in a given molecule does not vary with small changes from the equilibrium bond length, and $P(r_{eq})$, which relates the equilibrium bond length to the bond order in a series of molecules, is still fuzzy, the reader may demonstrate in a *reductio ad absurdum* proof that $P(r_{eq})$ must be differentiated when the derivative in (39) is taken. If the derivative of $P(r_{eq})$ is not taken, proceeding to (42) leads to the prediction that all vibrational force constants are zero!

[12] See the paper by Longuet-Higgins and Salem referred to at the end of this chapter.

From these two expressions A can be obtained in terms of the ratio k'/k as

$$A = \left(1 - \frac{k'}{k}\right)\frac{1}{0.18} \qquad (45)$$

which gives $A = 2.7 \text{ Å}^{-1}$.

We can use this value of A in (35) to obtain the required corrections to the calculated resonance energies due to the energy changes in the sigma system. We calculate from (35) that the energy required to compress a sigma bond from 1.52 Å to 1.43 Å, the calculated bond length in symmetrical cyclobutadiene, is $-0.08\beta_o$/bond. The sigma compression energy for symmetrical benzene ($r_{eq} = 1.40$ Å) is $-0.15\beta_o$/bond and for a pure double bond ($r_{eq} = 1.34$ Å) it is $-0.42\beta_o$. Thus, we calculate that the sigma skeleton of symmetrical cyclobutadiene is $-0.52\beta_o$ more stable than that of cyclobutadiene consisting of two double and two single bonds. Combined with the calculation that the π system of the symmetrical molecule is $-0.88\beta_o$ less stable, we conclude that cyclobutadiene should be $-0.36\beta_o$ less stable as a square than as a fully bond-alternated molecule. In benzene we find that the sigma system also contributes heavily to the calculated resonance energy. Here the calculation predicts that the sigma system of symmetrical benzene is more stable by $-0.36\beta_o$ than the fully bond alternated skeleton. Thus, we calculate the resonance energy of benzene—the difference in energy between the symmetrical and fully bond alternated forms, assuming no interaction between the double bonds in the latter—to be $-1.16\beta_o$, of which only about 70% comes from delocalization of the pi electrons.

Resonance Energy of Benzene and the Estimation of β_o

We have seen that thus far Hückel theory has proved to be quite successful qualitatively. Now we have calculated an actual number, the resonance energy of benzene. We predict that the experimental difference between the energy of benzene and that of the hypothetical cyclohexatriene will be $-1.16\beta_o$, and it is tempting to try to assess the quantitative accuracy

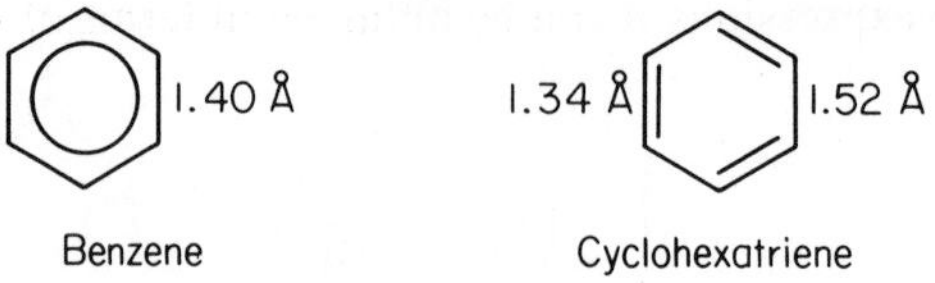

Figure 3.9

of the theory by comparing this theoretical value with an experimental figure.

Of course, since cyclohexatriene is a fictional molecule, we cannot measure the energy difference between it and benzene directly, but we will have to estimate it as best we can. Hydrogenation of benzene gives cyclohexane; hydrogenation of the fictional triene would give the same product. Both reactions would be exothermic; and we anticipate that since benzene is predicted to be the more stable of the two molecules, its hydrogenation should liberate less heat. In fact, the amount by which its hydrogenation is less exothermic than that of the fictional cyclohexatriene is exactly equal to its greater stability. Now we might try to approximate the heat of hydrogenation of each of the three di-substituted double bonds in cyclohexatriene by that of *cis*-2-butene, but how good an approximation is this?

In hydrogenating cyclohexatriene, not only would three double bonds between sp^2 carbons become bonds between sp^3 atoms, but three single bonds between sp^2 carbons also would be transformed into bonds between sp^3 atoms. In hydrogenating three molecules of *cis*-2-butene, the change for the double bonds is the same, but here six single bonds between sp^2 and sp^3 carbons are transformed into bonds between sp^3 atoms. Thus, the heat of hydrogenation of three *cis*-2-butene molecules, which is found to be -85.8 kcal/mole, will differ from that of our fictional molecule by $3[\Delta E(sp^2 - sp^2) - 2\Delta E(sp^2 - sp^3)]$, where ΔE is the *change* in energy in going from a single bond of the type in parentheses to one between sp^3 carbon atoms. Now this change in energy may be small and the quantity in brackets even smaller, compared to the heat of hydrogenation of 2-butene. Nevertheless, the empirical resonance energy of benzene is obtained by subtracting three times this latter quantity from -49.8 kcal/mole, the experimental heat of hydrogenation of benzene. Therefore, since we are taking the difference between two large numbers, the fact that one of them, our estimate of the heat of hydrogenation of cyclohexatriene, contains an error (namely, three times the quantity in brackets), may result in a large percentage error in the empirical resonance energy of benzene.

Problem 3.15. Show that the heat of hydrogenation of a fictional butadiene, consisting of two noninteracting double bonds, differs from that of two molecules of 1-butene by the quantity in brackets. The heat of hydrogenation of butadiene is found to be -57.1 kcal/mole, that of two molecules of 1-butene -60.6 kcal/mole. Thus, the resonance energy of butadiene is experimentally found to be 3.5 kcal/mole. Suppose that with a semiempirical value for β_0 the π resonance energy calculated in Problem 3.6 were very close to this value. Could you conclude that the quantity in brackets is necessarily zero? What other effect would you have to consider?

The uncertainty in the experimental value of the resonance energy of benzene arises from our use of a nonexistent molecule, cyclohexatriene, rather than a real one, as our reference structure. We might, therefore, consider adopting hexatriene as our reference structure, since it is a real molecule whose heat of hydrogenation can be measured experimentally and compared with that of benzene. However, this would only serve to transfer the uncertainty in the resonance energy of benzene, when redefined as the difference between its heat of hydrogenation and that of hexatriene, from experiment to theory. We would have to include in our calculated value of the resonance energy, thus redefined, the energy difference resulting from the difference in the changes that the sigma systems of the two molecules undergo on hydrogenation. Thus, adopting a real molecule, hexatriene, as our reference structure does not eliminate the problem encountered in comparing with experiment a theoretical value for the resonance energy of benzene. Such a change in the reference structure does mean, however, that the energy difference that we are computing is essentially that between three double bonds conjugated linearly (hexatriene) and cyclicly (benzene). This energy difference will not, of course, be the same as that between benzene and three isolated double bonds. Which reference structure and, hence, which definition one elects to use for the resonance energy of benzene is really just a matter of preference. We will continue to use the three unconjugated double bonds in the fictitious cyclohexatriene as the reference for quantitatively comparing a calculated resonance energy for benzene with an experimental value. Later in this chapter, however, we shall find it more convenient, in obtaining semiquantitative estimates of relative stabilities, to compare the pi systems of conjugated cyclic hydrocarbons with those of acyclic polyenes containing the same number of double bonds.[13]

Let us assume that $85.8 - 49.8 = 36$ kcal/mole does represent the energy difference between benzene and cyclohexatriene with which we want to compare our calculated figure. In order to make this comparison we need a value for β_0. Since we have already used force constant data to determine a value for the important parameter A, for the sake of consistency let us use the same data to obtain a value for β_0. Using $A = 2.7 \, \text{Å}^{-1}$ and $k = 7.62 \times 10^5$ dynes/cm $= 1.1 \times 10^4$ kcal/mole-Å^2 at $r_{eq} = 1.40 \, \text{Å}$ in (43) gives $\beta_0 = -43$ kcal/mole. This value for β_0 gives a calculated benzene resonance energy of 1.16×43 kcal/mole $= 50$ kcal/mole, almost forty percent too high. Does the fault lie with the Hückel method for pi systems, our treatment of sigma systems, or the way that we determined the value of β_0? While criticisms can be levied at all three, the major problem comes from the fact that we have parameterized our calculation using force constant data in order to calculate a quite different type of quantity, namely a resonance energy. Although we might hope that one set of parameters would allow us to calculate successfully a wide range of molecular properties, this wish is not always realized in practice; and it is generally best to choose key parameters like β_0 by fitting data of the same type that one wishes to calculate.

Problem 3.16.   A value of $\beta_0 = -3.0$ eV $= -69$ kcal/mole, obtained from the energy of the lowest electronic transition in ethylene, gives a poor fit to the experimental resonance energy of benzene. Are you surprised? Why?

---

[13] The use of acyclic polyenes as reference structures for computing resonance energies of cyclic molecules containing the same number of pi electrons does have an advantage, if in one's calculations the variation of β with bond length is ignored. The stability of both pi systems will be overestimated, but some cancellation of errors will occur when the difference in energies is taken. In contrast, since the stability of an isolated ethylene will not be overestimated, no such cancellation will occur if a molecule containing unconjugated double bonds is used as reference. For instance, if we ignore the variation of β with bond length (and, for consistency, the changes in the energy of the sigma bonds), in Hückel theory cyclobutadiene has the same energy as two ethylenes; hence, its resonance energy is zero. However, if we compare its pi energy to that of butadiene, the qualitatively correct prediction, that the cyclic molecule has a substantial negative resonance energy, is made.

Although in some cases it is obvious why a set of parameters that predicts well one type of molecular property fails miserably in predicting another, the reason why a value for β_0 from force constant data does so badly in predicting the resonance energy of benzene is somewhat more obscure. It arises from the fact that Hückel theory does not deal explicitly with electron repulsion. This causes little error in the treatment of the force constants in benzene, since they depend on the second derivative of the energy, to which electron repulsion terms make a relatively small contribution. Therefore, our value of β_0 is very close to the one we would use even if electron repulsion were included explicitly in the Hamiltonian. However, as we will see in Chapter 8, electron repulsion does contribute significantly to the resonance energy of benzene, since the electron repulsion in benzene is calculated to be about 16 kcal/mole greater than that in cyclohexatriene. Thus, when electron repulsion is included, the resonance energy of benzene is calculated to be $-1.16\beta_0 - 16$ kcal/mole. Therefore, a value of $\beta_0 = -43$ kcal/mole gives a theoretical value of 34 kcal/mole, in excellent agreement with the experimental number for the benzene resonance energy, but only when electron repulsion is included in the calculation, as it is in nature.

The failure of the above value of β_0 to give correctly the resonance energy of benzene provides a fine example of how in semiempirical calculations it is quite easy to get an incorrect answer for a theoretically significant reason, in this case because Hückel theory does not deal explicitly with electron repulsion. Such a failure is informative in showing us the limitations of an approximate theory and provides us with a motivation for improving it. On the other hand, it is also possible in semiempirical calculations to obtain the right answer for the wrong reasons; and such a success can delude one into thinking he is using a correct theoretical model, when that is not at all the case. For instance, if we were to omit the contribution of the sigma bonds to the resonance energy of benzene, we would calculate a value for this quantity of $-0.80 \beta_0$. Using $\beta_0 = -43$ kcal/mole, the benzene resonance energy would be calculated as 34 kcal/mole, in excellent agreement with experiment. We thus might be led to believe that the sigma bonds make a negligible contribution to the benzene resonance energy. Of course, this is not the case; in fact, we calculate that the difference in energy between the sigma bonds in benzene and those in cyclohexatriene contributes more than thirty percent of the net stabilization of the former over the latter. We could be led astray by the fortuitous cancellation of two effects—one due to the sigma bonds, which stabilizes benzene relative to cyclohexatriene, and the other due to electron repulsion, which has the opposite effect. Clearly, the fact that a calculation on a single molecule may give a value for a quantity that is in good agreement with experiment does not mean that the theoretical model on which the calculation is based is an accurate one.

Even if a theoretical method is successful in computing accurately one or more properties of a whole series of molecules, although the method may then be termed useful, the model on which it is based still may not be an accurate one. The success of calculations based on such a model may be due either to the fortuitous, but nevertheless systematic, cancellation of two effects that are ignored or to the ability of an adjustable parameter to encompass more than one effect. For example, were it generally the case that the difference in electron repulsion between two pi systems containing the same number of electrons, was directly proportional to the difference in their Hückel energies, then an equation of the form $\Delta E^{er} = k\Delta E^H$ would hold. Since the total energy difference, ΔE, is the sum of these two, $\Delta E = \Delta E^{er} + \Delta E^H = (k + 1)\Delta E^H$, so that the experimental energy difference would be proportional to that calculated by Hückel theory. Moreover, the factor $k + 1$ would automatically be included in β_0, if β_0 were chosen to fit an experimental energy difference. Thus, setting $\beta_0 = -36$ kcal/mole/1.16 = -31 kcal/mole might be a very good idea, provided we do something with this value of β_0 other than use it to show that Hückel theory gives a value of the resonance energy of benzene that agrees exactly with experiment.

We might, for instance, use this value of β_0 to predict that cyclobutadiene should be $-0.36\beta_0 = 11$ kcal/mole more stable in a rectangular than in a square geometry. Although electron repulsion does work to destabilize benzene relative to cyclohexatriene, as we shall see in Chapter 7 it works in the opposite direction for molecules like cyclobutadiene, tending to make the symmetrical geometry more stable than the bond-alternated one. Thus, electron repulsion serves to decrease the magnitude of the energy difference between symmetrical and bond-alternated geometries in both aromatic and antiaromatic molecules; but there is no assurance that this effect, which we will find has a different origin in the two series, can be included in the same value of β_0 for both types of compounds. In semiempirical methods like Hückel theory, because one parameter must often encompass several effects, success with a particular value of a parameter in calculating a property of a series of molecules does not guarantee success in another series. In fact, in the simplest versions of Hückel theory it is often necessary to choose a new value of β every time one wishes to calculate a different property of the same series of molecules.

A strong motivation for developing a more complete theoretical model is the discovery that no reasonable parameterization allows the prediction of a particular property or properties for a series of molecules. While Hückel theory can, in fact, be parameterized so that it can be used to successfully compute a wide variety of properties for certain classes of molecules,[14] in Chapter 5 we shall discuss two types of phenomena that unmodified Hückel theory is incapable of explaining. Hückel theory fails to account for these phenomena, because their origin lies in the fact that electrons do repel one another. In the next section we shall discuss briefly another failure of Hückel theory, which also results from its not dealing explicitly with electron repulsion.

Limitations to the Validity of the 4n + 2 Rule

We have found that by using HMO theory we can predict with some quantitative accuracy that monocyclic polyenes, or as they are sometimes called, annulenes, should be "aromatic," if they have $4n + 2 \pi$ electrons and "antiaromatic" if they contain $4n$ electrons. We defined an annulene as "aromatic" if it had all bond lengths equal and showed some special stability with respect to an appropriate reference compound with alternating single and double bonds. One trivial limitation to the validity of the $4n + 2$ rule will be encountered if we try to apply it, without modification, to systems other than the ones for which it was derived. For instance, if we uncritically apply the $4n + 2$ rule to any system for which a resonance structure can be written that places $4n + 2$ electrons around the periphery of a ring, we are bound to make incorrect predictions. In the case of naphthalene two resonance structures can be written which place 10 π electrons around the periphery; however, bond alternation is observed in this molecule. This is not so surprising since the third resonance structure would tend to shorten the 1-2 relative to the 2-3 bond. Experimentally the former is about 0.05 Å shorter than the latter. In anthracene, as well, the 0.04 Å shortening of the 1-2 relative to the 2-3 bond is explicable on the basis of contributing resonance structures which do not place 14 π electrons around the periphery.

---

[14] See Streitweiser's book in the suggestions for further reading at the end of the chapter.

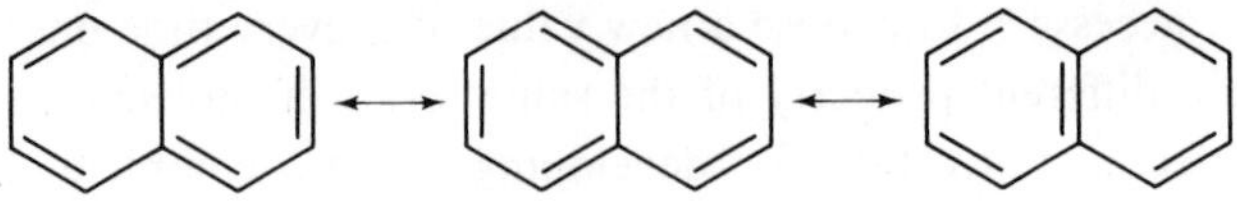

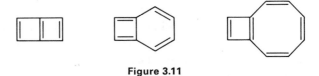

Figure 3.10

In MO theory the different bond lengths can be understood in terms of perturbations of the MO's in the corresponding annulene by the σ bridges, which permit interactions, not present in [10] and [14]annulene, between AO's in naphthalene and anthracene. Yet, if we slavishly apply the $4n + 2$ rule and the definition of aromaticity, we overlook these perturbations. In extending the $4n + 2$ rule, or any other theoretical result, to a system for which it was not explicitly derived, one must be careful to assess the changes to be expected from the perturbations present in the new system.

Problem 3.17.   Write resonance structures for the molecules shown in Figure 3.11. Use the $4n + 2$ rule critically to assess which will contribute most heavily. Which bonds do you expect to be longest in each molecule?

Figure 3.11

A very subtle type of perturbation is that the reference theory from which a rule is derived may no longer work in a new system. For example, both the cyclopropenium cation and cyclopentadienyl anion enjoy considerable resonance stabilization. On joining the two together to form pentatriafulvalene, one might expect from the $4n + 2$ rule that the molecule would resemble **A** more than **B**; and an HMO calculation bears out this expectation. However, we should remember that electrostatic effects are not explicitly included in HMO calculations so that HMO theory will surely overestimate the amount of charge separation. In this case Hückel theory predicts a dipole moment of nearly 14D, while a method discussed in Chapter 8, which explicit accounts for electrostatic effects in the Hamiltonian, reduces this to little more than 3D.

More interesting theoretically than limitations to the validity of the $4n + 2$ rule, due to its application to systems for which it was not derived,

A   B

Figure 3.12

is the question of whether bond alternation may be found, even in annulenes containing $4n + 2$ π electrons, for large values of n. Let us anticipate why bond alternation might occur. From the circle mnemonic for (23) we can see that as N, the number of atoms in the ring, becomes larger, the gap between bonding and antibonding energy levels becomes smaller. Therefore, if a small distortion in the geometry occurs, which allows the mixing of unfilled antibonding orbitals with filled bonding levels, the energy lowering due to this mixing will be biggest in large rings where these levels are closest together. When the π energy lowering, due to the orbital mixing on molecular distortion, is larger than the increase caused by changes in the σ compression energy, a symmetrical configuration will represent a saddle point of the potential energy surface and will thus be unstable. We can put this mathematically by saying that so long as the second derivative of the total energy, E^T, of a molecule, evaluated in the symmetrical configuration, remains positive, the symmetrical configuration remains an energy minimum; but when

$$\frac{\delta^2 E^T}{\delta r^2} = \frac{\delta^2 E^T_\sigma}{\delta r^2} + \frac{\delta^2 E^T_\pi}{\delta r^2} \qquad (r = r_{\text{sym}}) \tag{46}$$

becomes negative, the molecule distorts. In our investigation of the possibility of bond alternation in a large enough annulene through a "pseudo Jahn-Teller" distortion, we will follow the approach of Longuet-Higgins and Salem and examine the behavior of the second derivative of the total energy for very large N.

In order to obtain an expression for the effect on $\delta^2 E^T/\delta r^2$ of increasing N, we need to obtain a more general expression for the π electron energy; for in the event that the molecule distorts so that bond lengths alternate, the use of two different values of β becomes necessary. We proceed as before, but now one of our MO's has the form

$$\psi = \sum_{r=1}^{N/2} \phi_{2r-1} \cos 2\omega r + \phi_{2r} \cos 2\omega(r + \chi) \tag{47}$$

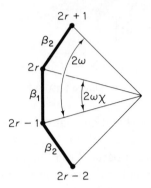

Phase Relations between AO's in a
Bond–Alternated Annulene

Figure 3.13

instead of that implied by (19), because as shown in Figure 3.13, every other atom is shifted by a constant phase factor $2\omega\chi$ from the one before it by bond alternation.

Proceeding as before, we obtain, on multiplying through by ϕ_{2r-1} and integrating,

$$E - \alpha = \frac{\beta_1 \cos 2\omega(r + \chi) + \beta_2 \cos 2\omega(r - 1 + \chi)}{\cos 2\omega r} \tag{48}$$

and, after multiplication by ϕ_{2r} and integration,

$$E - \alpha = \frac{\beta_1 \cos 2\omega r + \beta_2 \cos 2\omega(r + 1)}{\cos 2\omega(r + \chi)} \tag{49}$$

The trigonometric formulas involved in the solution of these two equations lead to rather cumbersome expressions; and it is much easier to use complex phase factors $e^{i\phi} = \cos\phi + i\sin\phi$ in place of the cosines in (47–49). Making this substitution and multiplying together (48) and (49) gives

$$(E - \alpha)^2 = (\beta_1 e^{2i\omega\chi} + \beta_2 e^{2i\omega(\chi-1)})(\beta_1 e^{-2i\omega\chi} + \beta_2 e^{2i\omega(1-\chi)}) \tag{50}$$

which reduces to

$$(E - \alpha)^2 = \beta_1^2 + \beta_2^2 + 2\beta_1\beta_2 \cos 2\frac{2\pi m}{N} \tag{51}$$

Like (23), to which it reduces for $\beta_1 = \beta_2$, (51) has a simple geometric interpretation. It is just the law of cosines formula for the length, $E - \alpha$, of the third side of a triangle when the other two sides are of length β_1 and β_2 with an angle of 2ω between them, where ω can only assume the values $2\pi m/N$ ($m = 0, 1, 2, \ldots N/2$). The third side of a triangle must be smaller in magnitude than the sum of the other two and larger than their difference; therefore, for the negative root of (51) $\beta_1 + \beta_2 < E - \alpha < \beta_1 - \beta_2$, assuming $\beta_1 < \beta_2$, and for the positive root, $-(\beta_1 + \beta_2) > E - \alpha > \beta_2 - \beta_1$. Thus, unlike the fully symmetric system, when the bonds alternate the energy levels fall into two distinct bands with a separation of $2|\beta_2 - \beta_1|$ between the bonding and antibonding levels.

We can now write the total energy as

$$E^T = NE_\sigma + N\alpha - 2\left(\sum_{m=(N-2)/4}^{-(N-2)/4} \sqrt{\beta_1^2 + \beta_2^2 + 2\beta_1\beta_2 \cos \frac{4\pi m}{N}}\right) \quad (52)$$

where the total σ energy E_σ^T has been written as N times that for one σ bond, and the summation spans the filled π orbitals in a $4n + 2$ system. For some distortion which shortens by x all the bonds with which we have associated β_1 and lengthens the alternating bonds by the same amount, we can write $\beta_1 = \beta_s e^{+Ax}$ and $\beta_2 = \beta_s e^{-Ax}$ where β_s is the value of β for the symmetric configuration. The second derivative of the total energy with respect to this distortion is then

$$\frac{\delta^2 E^T}{\delta x^2} = (4n + 2)\frac{\delta^2 E_\sigma}{\delta x^2}$$

$$- 2 \sum_{m=n}^{-n} \left(\frac{\delta^2}{\delta x^2} \sqrt{\beta_1^2 + \beta_2^2 + 2\beta_1\beta_2 \cos \frac{2\pi m}{2n + 1}}\right) \quad (53)$$

where we have made the substitution $N = 4n + 2$.

If the second term in (53) increases faster with n than the first, so that for some value of n $(\delta^2 E^T/\delta x^2)$ becomes negative, the equilibrium becomes unstable and distortion will occur. We can still, however, use the expression (41) for the second derivative of the sigma energy of a bond at equilibrium. Employing (11) and (12), (41) can be rewritten

$$\left(\frac{\delta^2 E_\sigma}{\delta x^2}\right)_{x=0} = -2A^2\beta_s\left(\frac{1}{0.18A} + P(r_{eq})\right) \quad (54)$$

The quantity in parentheses, using $A = 2.7$ Å^{-1} and $P(r_{eq}) = 0.67$ for an aromatic compound, is 2.74.

The second derivative of the π energy term, evaluated at $x = 0$, is

$$\left(\frac{\delta^2 E_\pi^T}{\delta x^2}\right)_{x=0} = 2 \sum_{m=n}^{-n} 2\beta_s A^2 \sec\left(\frac{m\pi}{2n+1}\right)$$

$$= 4\beta_s A^2 \sum_{m=n}^{-n} \sec\left(\frac{m\pi}{2n+1}\right) \tag{55}$$

which is negative and diverges like $n[\ln(n)]$ for large values of n. Thus, no matter what the exact form of $\delta^2 E_\sigma/\delta x^2$, for large enough values of n the π term must certainly be greater in magnitude, and so the molecule will distort. Using our expression for $\delta^2 E_\sigma/\delta x^2$ from (54) in (53) and replacing the second term of (53) by (55), we find that the value of n at which distortion takes place is given implicitly by the expression

$$\frac{1}{2n+1} \sum_{m=n}^{-n} \sec\left(\frac{m\pi}{2n+1}\right) > 2.74 \tag{56}$$

For $n = 8$ the left hand side of (56) has the value 2.766 and for $n = 7$, 2.687. Thus, we would predict that [34]annulene should be the first to show bond alternation. Experimentally, Sondheimer has found that the nmr spectrum of tridehydro[26]annulene does not give evidence of the diamagnetic ring current seen in the deshielding of the ring protons in lower [4n + 2]-annulenes.[15] However, monodehydro[26]annulene does have a ring current. This is taken as indicating that bond alternation has not yet set in but that a sufficient number of short acetylenic bonds can induce alternation at $n = 6$. Our theory thus seems at least semiquantitatively correct. But even should alternation set in at $n = 7$ instead of 8, this ought not to dismay us; for the prediction of the exact value of n for the onset of bond alternation is extremely sensitive to A. If A were, in fact, 2.8 instead of 2.7 Å^{-1}, the theory would predict the onset of alternation at $n = 7$.

Perturbation Treatment for Large π Systems

We have now seen that HMO theory is not only in excellent agreement with experiment regarding the qualitative stability, or lack of it, of many conjugated cyclic systems, but that it can also be parameterized to

---

[15] For a review of the annulenes see *Accts. Chem. Res.*, **5**, 81 (1972).

make quantitative predictions of resonance energies and of the onset of bond alternation in $4n + 2$ annulenes. As we have formulated it, however, HMO theory suffers from the drawback that every time we wish to make a prediction, we have to carry out a calculation. This can require considerable time for a molecule the size of naphthalene, even when use is made of its high symmetry.[16] A molecule of lower symmetry like azulene really requires a computer.[17] The nice thing about resonance theory is that with pencil and paper one can in a few minutes draw some resonance structures and make a prediction. Of course, the predictions are not always correct; for example, cyclobutadiene *is* less stable than benzene, despite the fact that two equivalent resonance structures can be drawn for each. Therefore, it would be extremely convenient if HMO theory could be formulated in such a way that with no more time and effort than it took to draw resonance structures, one could make qualitative or even semiquantitative predictions.

While it is true that the $4n + 2$ rule does enable one to make qualitative predictions regarding stability of π systems, we have already noted that its strict validity is really limited to the small annulenes. The $4n + 2$ rule can be combined with resonance theory, as we have done in Problem 3.17, to make predictions about polycyclic systems, by considering that those resonance structures which generate $4n + 2$ electron cycles and avoid those of $4n$ electrons contribute most heavily. However, the question still arises as to the relative contributions of two structures, one of which places 6 π electrons around a six membered ring, the other of which destroys the benzenoid ring but places $4n + 2$ electrons around a larger periphery. This is a general type of problem with resonance theory—too much is left to the subjective judgement of the user. It would be gratifying to have an MO theory that was completely self-contained and did not rely on resonance theory at all, yet was as simple to apply to large π systems.

Such a treatment can be developed using perturbation theory, symmetry, where applicable, and the results of calculations on small systems. The strategy is to calculate the properties of a large π system from those of smaller systems into which it can be formally dissected, using perturbation

---

[16] For some hints on setting up and solving by hand the HMO equations for large systems see Chapter 2 of Streitweiser's book, which is suggested at the end of this chapter for further reading.

[17] The results of simple HMO calculations on the π systems of a large number of organic molecules are tabulated by C. A. Coulson and A. Streitweiser in *Dictionary of π Electron Calculations*, W. H. Freeman, San Francisco, 1965.

theory to find the changes caused by the interaction between the constituents. For example, cyclobutadiene can be thought of as consisting of two ethylenic π systems. By exploring the consequences of their interaction, we ought to be able to deduce the properties of cyclobutadiene. Now the bonding π MO of one ethylenic unit will not interact with the antibonding MO of the other since they are of different symmetry. The only interactions will be between the two bonding MO's and between the two antibonding MO's. Because the MO's start off degenerate in energy, we cannot use Equation (23) of Chapter 2 to find the new MO's nor Equation (24) to find the new energies. In fact, we need no formulas to solve this problem, for we already know that when two wavefunctions of the same energy are mixed by a perturbation, $\mathcal{3C}'$, one is lowered by the interaction energy, $\int \psi_i \mathcal{3C}' \psi_j \, d\tau$, corresponding to the equal in-phase mixing of the two original wavefunctions, and the other is raised by the same amount, corresponding to out-of-phase mixing. We also know the coefficients in ethylene all to be of magnitude $1/\sqrt{2}$. Therefore, the energy of interaction between the MO's will be $(2 \times 1/\sqrt{2} \times 1/\sqrt{2})\beta = \beta$, the factor of 2 being necessary because the ethylenes interact at both ends of each. Thus, the mixing will result in a drop of β for one combination of the interacting MO's and an equal increase in energy for the other.[18] The results are summarized in Figure 3.14, and the reader may check that the MO's as well as the MO energies of cyclobutadiene are reproduced. Moreover,

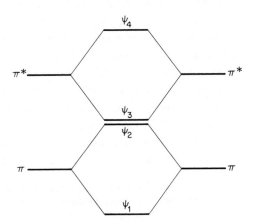

Orbital Interaction Diagram for Cyclobutadiene,
Considered as Two Ethylenic π Systems

Figure 3.14

since the mixing which results in a drop in energy of an MO that originally contained two electrons is accompanied by an equal rise in energy of an MO that also contained two electrons, no net stabilization is achieved. Consequently, we predict no stability for cyclobutadiene over two isolated ethylenic units.[18]

It follows generally that since one orbital can lower its energy only at the expense of raising that of another, interactions between filled orbitals will change orbital energies but not total energies. The corollary of this is that net stabilization of a system can only be achieved through interaction between filled, or partly filled, levels in one component and unfilled, or partly filled, levels in the other.

Problem 3.18. Why will interaction between two orbitals, each *initially* containing one electron, lead to net lowering of the energy of a system?

An example of a system where net stabilization occurs through mixing of filled with empty orbitals is benzene, when it is analyzed as being formed from the π systems of ethylene and butadiene. The bonding MO in ethylene has the proper symmetry for interaction with both ψ_1 and ψ_3 of butadiene. The interaction with the former destabilizes the bonding ethylene MO, that with the latter stabilizes it. Apparently the two opposite effects cancel, for we note that there is an MO left at energy $\alpha + \beta$ in benzene. The net result of these two interactions is that an unfilled orbital, ψ_3, increases in energy, while a filled level, ψ_1, decreases.[19] In similar fashion the antibonding ethylene MO can mix with ψ_2 and ψ_4 of butadiene. The important effect for the total energy of the molecule is the mixing of π^* with ψ_2, which is filled. Any raising of the antibonding ethylene MO which results does not affect the total energy, since π^* is empty. Actually, the raising of π^*, caused by mixing with ψ_2, is cancelled by the interaction

---

[18] This, of course, is only true because we are neglecting overlap. When it is explicitly included in the calculation, the decrease in energy of the in-phase combination is less than the increase in that of the out-of-phase combination. When overlap is included, cyclobutadiene, even after distortion to a rectangle, is less stable than two ethylenes.

[19] The statement that "ψ_1 and ψ_3 mix through their interaction with the ethylene bonding MO" could be somewhat misleading in that it might be taken to imply that the latter remains unaffected. This is hardly the case; for while the energy of this MO is unchanged after mixing, it has been transformed into a benzene MO which bears only partial resemblance to the original bonding ethylene orbital.

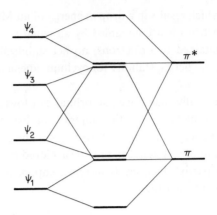

Orbital Interaction Diagram for Benzene, Considered as Being
Formed from the π Systems of Ethylene and Butadiene

Figure 3.15

of π^* with ψ_4. The orbital interaction diagram for the formation of benzene from ethylene and butadiene is shown in Figure 3.15.

Problem 3.19. Use second order perturbation theory [Equation (24) of Chapter 2] and the tabulated MO's for butadiene to determine the energies of the benzene MO's resulting from the mixing of ψ_1 and ψ_3 with the ethylene bonding MO.

The problem with the foregoing treatment of benzene is that while it does tell us that benzene can be expected to be more stable than the isolated π systems of butadiene and ethylene, and even how much more stable, if we take the time to actually do the calculation, it still requires more work than we might wish. First, it requires that we know, look up, or calculate the MO's of butadiene. For a larger system this could prove much more of a problem. Second, we have to take into account all the interactions between MO's of the same symmetry. We might have been tempted to ignore the interaction between the ethylene bonding MO and ψ_3 because their energy difference of 1.62β is greater than that of 0.62β between the former and ψ_1. However, experience (Problem 3.19) showed us that this would have been a mistake; the changes in orbital energies caused by the two interactions turn out to be the same because of the larger coefficients at the terminal atoms in ψ_3. Finally, in order to compute the effect of mixing on the energies, we did have to do a little arithmetic.

Clearly, in this perturbational approach we have not reached the simplicity of resonance theory yet.

We should like to construct our perturbational method so it would enable us to confine our attention to just two interacting MO's. Ideally, they should be chosen so we could write them down at a glance and would turn out to be degenerate so that once we had calculated their interaction energy, $\int \psi_i \mathcal{JC}' \psi_j \, d\tau$, the problem would be solved. Now what we would like to do with such a theory is predict differences in total energies between π systems; therefore, the one interaction term that we compute must contain most of the net stabilization energy on joining two π systems. All other interactions must, hopefully, be small compared to it. Clearly then, it cannot be an interaction between filled MO's, for this leads to no net stabilization. Nor can it be an interaction between an unfilled and a filled MO, for there are usually several of these of about the same magnitude in a system; and anyway, we would prefer our one interaction to be between MO's that are degenerate in energy. This will not only make the energy change caused by their interaction large but will also make our computational work small. The only possibility left is to dissect a system in which we are interested into components, each of which has one non-bonding MO (NBMO). Provided we can easily obtain the required coefficients, we have just what we want—a method of analysis in terms of one interaction, between degenerate MO's, which should contain most of the net energy change on fusion of the two components.

As an example, let us return to benzene and analyze the π system as the interaction of two allyl fragments, each of which has an NBMO. Now although the two allyl ψ_1 MO's will interact strongly, they are both filled so that their mixing will lead to no net stabilization. Therefore, the only net stabilization will arise from interaction of the two NBMO's, apart from whatever stabilization comes from the interaction of the filled ψ_1 in one fragment with the empty ψ_3 in the other. This latter interaction is expected to be small compared with that between the NBMO's because of the large energy difference between ψ_1 and ψ_3 and also because in ψ_2 the coefficients at the termini are larger than those in ψ_1 or ψ_3. Of course, ψ_2 does not interact with either ψ_1 or ψ_3 because of the difference in symmetry. The interaction diagram is shown in Figure 3.16. From the coefficients for ψ_2 in allyl we calculate that the in-phase combinations of NBMO's will be lowered in energy by $(2 \times 1/\sqrt{2} \times 1/\sqrt{2})\beta = \beta$. Since two electrons are to go in this MO we calculate that the π system of benzene will be more stable than that of two isolated allyl radicals by 2β.

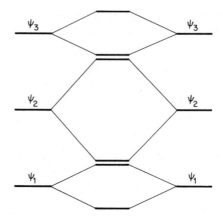

Orbital Interaction Diagram for Benzene, Considered
as Two Allylic π Systems

Figure 3.16

Unfortunately, this is not a very helpful piece of information because we have no clear idea of what it means. Since all we have is a vague feeling that two allyl radicals are not very stable at all, the prediction that benzene is going to be considerably more stable is not a very useful one. In other words, two allyl fragments are not a very useful reference system for our calculation. However, two allyl fragments can also be combined to make hexatriene. By just computing the energy of interaction between one end each of our two allyl systems, we find that the in-phase mixture of NBMO's will be lowered by $(1/\sqrt{2} \times 1/\sqrt{2})\beta = \tfrac{1}{2}\beta$, so that hexatriene is calculated to be stabilized by β relative to two allyls. Thus, benzene is predicted to be β more stable than hexatriene. This, needless to say, is at best a semiquantitative result; for we have not corrected our β's for bond length differences (nor can we easily do so using this approach). In addition, we have ignored the stabilizing interactions between bonding and antibonding orbitals, which, although they may be small compared to that between the NBMO's, still do exist and contribute to the energy difference. Nevertheless, without ever directly calculating the MO's of benzene or hexatriene or their energies, we have been able to show, with a minimum amount of work, that the π system of the former should be considerably more stable than that of the latter.

Problem 3.20.   Calculate the energy of the π system of cyclobutadiene relative to butadiene by considering the interaction of a lone p orbital with allyl.

We have now almost accomplished our objective of providing a theoretical method for evaluating reliably, within the MO formalism, the relative energies of π systems with the same "back of the envelope" ease provided by the less satisfactory resonance theory. What we still require is a method for rapidly computing the coefficients in the NBMO's of the fragments into which we dissect a π system. For example, if we were to wish to compare the stability of benzene and hexatriene by dissecting the π systems into pentadienyl and a carbon atom, we are clearly at no advantage to standard HMO theory if we have to carry out the usual Hückel calculation to find the coefficients in the NBMO of the former fragment. We shall now see that there is no necessity to do this; it is possible to find the required coefficients in less time than it takes to draw all the resonance structures for the system. To see how this can be done we now turn to a remarkable series of theorems concerning a class of molecules known as alternant hydrocarbons.

Alternant Hydrocarbons

An alternant hydrocarbon (AH) is defined as one in which it is possible to divide the carbon atoms into two sets, such that no two atoms belonging to the same set are joined by a bond. This implies that an AH can contain no odd-membered rings. Traditionally the atoms of the different sets are distinguished by placing a star next to all those of one set.

Figure 3.17

The reader may have noticed that in the alternants we have already discussed, some of which are sketched in Figure 3.17, the bonding and antibonding orbitals come in pairs, symmetrically distributed about $E = \alpha$, with the same coefficients, except for a sign change at all starred atoms. We can easily prove that the above pairing theorem should hold for all alternant hydrocarbons. We obtain the energy of an MO in the usual way by multiplying the Schrödinger equation by ϕ_r and integrating. If ϕ_r

is an unstarred atom then all those bonded to it are starred. We thus obtain

$$(E - \alpha)c_{ir} + \sum_s c_{is}^* \beta = 0 \qquad (57)$$

If ϕ_t, a starred atom, is used we get

$$(E - \alpha)c_{it}^* + \sum_u c_{iu} \beta = 0 \qquad (58)$$

where the sums are over just those atoms bound to r and t and the stars denote coefficients of starred atoms. Suppose we know the coefficients that correspond to a solution of (57) with $E - \alpha = k\beta$, Can we immediately find another solution? Let us change the sign of all the coefficients of one set of atoms in the wavefunction.

$$(E - \alpha)c_{ir} - \sum_s c_{is}^* \beta = 0 \qquad (59)$$

Clearly, if $E - \alpha = k\beta$ satisfies (57), Equation (59) will be satisfied by $E - \alpha = -k\beta$. Therefore, if a solution of (57) with $E - \alpha = k\beta$ and coefficients c_{is}^* exists, there is another with $E - \alpha = -k\beta$ and coefficients $-c_{is}^*$. Thus we have proven our first AH theorem: (1) In an AH each bonding orbital of $E = \alpha + k\beta$ is accompanied by an antibonding orbital of energy $E = \alpha - k\beta$ which differs from it only by a change in sign of the coefficient at each atom of one set (starred or unstarred). It follows immediately from the relationship between the coefficients in MO's which are so paired: (2) In MO's that are paired with each other, the charge densities are identical and the bond orders are equal in magnitude, but, between atoms belonging to different sets, opposite in sign. It is also possible to prove two more general theorems (See Appendix) about uncharged AH's: (3) In a neutral AH the π electron density at each atom is unity. (4) In a neutral AH the bond order between any two atoms of the same set (starred or unstarred) is zero.

    The most important use of the pairing theorem and its consequences for us at this point is its application to the π systems of alternant hydrocarbons consisting of an odd number of carbon atoms (odd AH's). Since the number of MO's is always equal to the number of AO's from which they are formed, in such a system the number of MO's must be odd. But according to the pairing theorem, each MO of energy $\alpha + k\beta$ has a partner of energy $\alpha - k\beta$; therefore, in an odd AH at least (and usually only) one orbital must be paired with itself and have energy α. Therefore: (5) In an odd AH there must be at least one NBMO. With $E = \alpha$, (57) requires that the coefficients of the starred atoms about an unstarred atom sum to zero, and (58) makes a similar demand on the coefficients of the unstarred atoms about a starred atom. Thus far we have not specified how to distinguish starred from unstarred atoms. We now adopt the convention that the set containing the most atoms will be starred. If, as is usually the case, an odd AH contains a chain with a starred atom at the end, the one atom next to it is unstarred. Equation (58) implies then that this unstarred atom's coefficient must be identically zero. In fact, it can be shown that the coefficients of all the unstarred atoms are identically zero in an NBMO. The coefficients of the starred atoms cannot also all be zero, or the NBMO vanishes. Thus: (6) The coefficients of the starred atoms about each unstarred atom must sum to zero in the NBMO of an odd AH,[20] while those of all unstarred atoms are identically zero.

We can now rapidly find the coefficients in the NBMO of an odd AH. It is usually convenient to start at a terminus and star every other atom. After making sure that the atoms of the starred set are the more numerous, the unstarred atoms can be assigned coefficient zero in the NBMO. We now arbitrarily assign one of the starred atoms a coefficient a. Experience shows that if there is a ring present, it is best to start with a starred ring atom that has no more than two other starred atoms as next nearest neighbors; or it may be necessary to assign fractional values of a to other atoms as we proceed. We then continue assigning $\pm a$ or a multiple to all the other starred atoms so that the sum of the coefficients of the starred atoms about any unstarred atom is zero. This gives the relative magnitudes of the coefficients in a NBMO; their absolute values can be obtained from normalization, since the sum of squares of coefficients must equal one. Some examples are shown in Figure 3.18. The arrow indicates the starting point for each molecule.

Problem 3.21.   Calculate the coefficients in the NBMO for the systems in Figure 3.19.

---

[20] It may sometimes occur that the atoms of a molecule containing several termini cannot be starred in such a way that all the terminal atoms are starred. In chains with an unstarred terminal atom, the fact that such an atom can have only one starred neighbor in conjunction with theorem (6) requires that the coefficients of all the atoms, unstarred and starred, in that chain be zero in the NBMO.

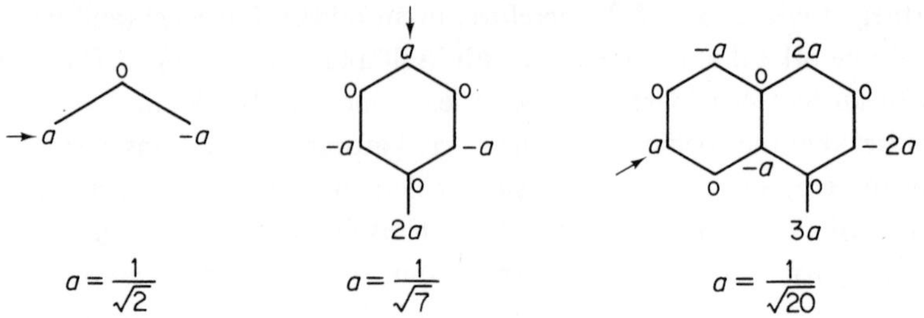

NBMO's in Some Odd AH's

Figure 3.18

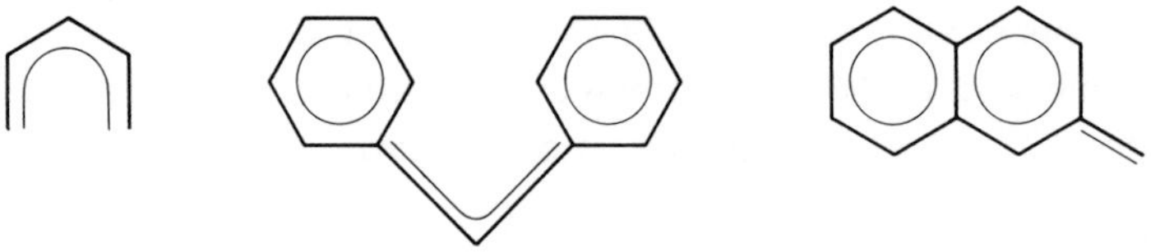

Figure 3.19

Problem 3.22.   Calculate the relative stabilities of benzene and hexatriene by considering the interaction between pentadienyl and a single carbon atom.

Applications of PMO Theory

The ease of computing coefficients in an NBMO makes the perturbational MO(PMO) approach ideally suited to rapidly estimating the relative energies of π systems. This method, developed by Dewar, even enables us to actually derive the $4n + 2$ rule for stability.

An odd AH must consist of either $4n + 1$ or $4n - 1$ carbon atoms. The number of starred atoms in each is $2n + 1$ and $2n$ respectively. Since in a chain the coefficients of the starred atoms in the NBMO are all of the same magnitude and alternate in sign, the terminal atoms in a $4n + 1$ chain will have the same sign and, in a $4n - 1$ chain, opposite signs. The former system can then interact at both termini with an additional carbon atom to give a $4n + 2$ cyclic system with a net energy lowering of $2^{21} \times 2a\beta$, or it can form a linear system with energy lowering of $2 \times a\beta$. Thus,

---

[21] The factor of 2 arises from the fact that two electrons go into the in-phase combination of NBMO's.

a cyclic $4n + 2$ system will be $2a\beta$ more stable than the corresponding chain. For a $4n$ system, there is no net interaction between the ends of the $4n - 1$ chain and a lone carbon atom, because of the opposite coefficients of the terminal atoms in the NBMO. Thus, the cyclic $4n$ system is predicted to be $-2a\beta$ *less* stable than the corresponding open chain. Moreover, since the magnitude of a decreases with chain length, the effects of the $4n + 2$ rule should be manifested most strongly in small systems, for example, cyclobutadiene and benzene.

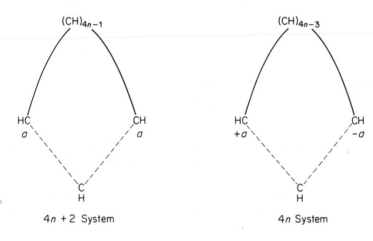

PMO Analysis of $4n + 2$ and $4n$ π Systems

Figure 3.20

Of course, the preceding is a derivation of just part of the $4n + 2$ rule, for it only applies to systems of $4n$ or $4n + 2$ atoms; whereas, the $4n + 2$ rule for stability also applies to ionic π systems in rings of $4n \pm 1$ atoms. Obviously we cannot treat these latter systems directly using the rules for odd AH's, since rings containing an odd number of atoms are not AH's. However, we can analyze them by imagining that these π systems are formed by joining the ends of the corresponding odd AH chains and then using first order perturbation theory to compute the change in energy. Perturbation theory tells us that the first order change in energy on joining the ends to form a ring is equal to $2p_{rs}\beta$, where p_{rs} is the net bond order between the terminal atoms in the open chain odd AH. Theorem (4) for AH's is relevent here since it states that the net bond order between the termini in the *neutral* open chain system is zero, because the terminal atoms both belong to the starred set. However, if we add another electron to the NBMO to form the anion, the bond order will no longer be zero. This additional electron will contribute $\pm a^2$ to the bond order, where a is the magnitude of the terminal atoms' coefficients in the NBMO, and the sign is determined by whether the signs of the coefficients are the same ($4n + 1$ atom chain) or opposite ($4n - 1$ chain). Thus, anions in $4n + 1$ atom rings will be more stable than in the corresponding open chain systems by $2a^2\beta$ and less stable by the same amount in rings of $4n - 1$ atoms. Since a neutral $4n + 1$ atom system contains $4n + 1$ π electrons and a $4n - 1$ system $4n - 1$ π electons, and since in forming an anion we are adding an electron to each of these, we again wind up with a restatement of the $4n + 2$ rule for anionic π systems.

Problem 3.23. Make arguments similar to the above for cations and show that the $4n + 2$ rule is again obtained.

Problem 3.24. What are the net charges at each atom in the pentadienyl radical? in the anion? [Hint: Use the AH theorems.] What are the charges at each atom in the cyclopentadienyl anion? [Use symmetry.] What does first order perturbation theory predict will be the charge distribution in the cyclic system formed by closure of pentadienyl anion? If there is a discrepancy between the perturbation and symmetry derived results, why does it exist?

We have now seen that we can very nicely derive the $4n + 2$ rule through the use of PMO theory. However, one of our purposes in developing this approach to π electron theory was to go beyond the $4n + 2$ rule for monocyclic π systems and predict the relative stabilities of polycyclics. We shall now show that the PMO approach can do this quite nicely for bicyclic molecules. Consider the three shown in Figure 3.21. Each can be dissected into a nonatetraenyl fragment and a single carbon atom by formally removing a bridgehead atom, as indicated. The coefficients in the NBMO of the nonatetraenyl fragment are also shown. In the first molecule the interaction between the ends of the chain and the carbon atom in forming [10]annulene results in an energy lowering of $(2 \times 2/\sqrt{5})\beta$, which is offset partially by the antibonding interaction in the bridge, resulting in a net energy relative to the two radical fragments of $(2 \times 1/\sqrt{5})\beta$. In azulene this antibonding interaction is absent, and azulene has the same π energy as [10]annulene. However, naphthalene is the most stable of all, because of the bonding interaction across the bridge,

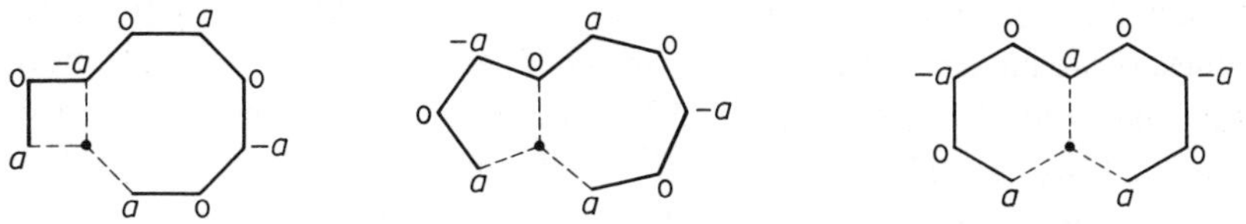

$$a = \frac{1}{\sqrt{5}}$$

PMO Analysis of the π Systems of $C_{10}H_8$ Isomers

Figure 3.21

giving it a relative energy of $(2 \times 3/\sqrt{5})\beta$. Experimentally, it is found that azulene rearranges to naphthalene, and derivatives of the first molecule behave like rather unstable polyenes.

Problem 3.25. What would you predict about the length of the bond bridging the ten-membered ring in bicyclo[6.2.0]decapentaene, the first molecule shown in Figure 3.21?

Now there is another way we could have used the properties of AH's to carry out the analysis of the relative energies of $C_{10}H_8$ isomers. We note that they all can be formed by joining two atoms in [10]annulene. We also know that the bond order between two atoms of the same set is zero in a neutral AH; and using first order perturbation theory, we can see that azulene, being formed from [10]annulene by joining two such atoms, should have the same π energy as [10]annulene. This is, of course, the same conclusion that we reached previously, based on the mode of analysis shown in Figure 3.21. However, because there is no simple way of determining the bond order between two atoms of different sets in an AH, without dividing it into odd AH fragments, this second approach does not distinguish between the first and third molecules as to which is lower in energy.

The same problem arises in trying to apply perturbation theory to [14]annulene in order to decide whether the π system of anthracene or phenanthrene is more stable. Both molecules can be analyzed as involving two interactions between atoms of different sets in [14]annulene; but since we don't know the relevant bond orders, we cannot proceed further. However, the analysis of relative stability can be carried out by dividing the two systems into *identical* odd AH fragments and comparing the energy lowering on rejoining the fragments. Obviously, the fragments into which the two molecules are dissected must be the same for both molecules, or we have no way of comparing energies. This is not meant to imply that the molecules must be divided into benzyl fragments; on the contrary, the analysis can also be made for β-allylnaphthyl interacting with a carbon and gives the same qualitative result. However, it is somewhat easier to find the NBMO in benzyl, and the analysis along this line is shown in Figure 3.22. It is found that the π system of phenanthrene is predicted to be $(2 \times 1/7)\beta$ lower in energy than that of anthracene.

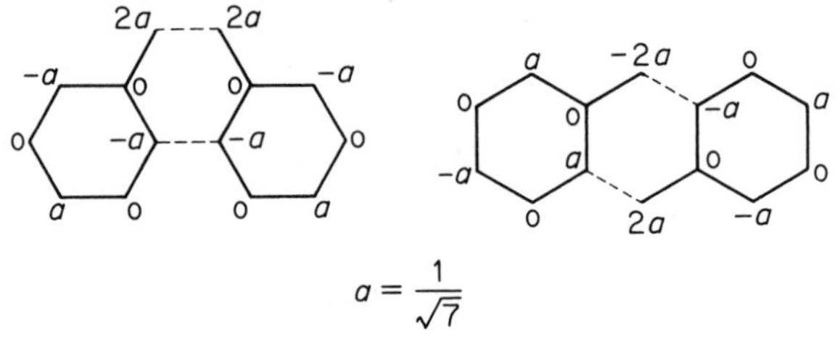

$$a = \frac{1}{\sqrt{7}}$$

PMO Analysis of the π Systems of Phenanthrene and Anthracene

Figure 3.22

Problem 3.26.   Carry out an analysis of the phenanthrene-anthracene problem by dissecting each into β-allylnaphthalene and a carbon atom as shown in Figure 3.23.

Figure 3.23

Although AH theorem (4), regarding the vanishing of bond orders, is not useful for comparing energies of molecules which have rings containing only even numbers of atoms, it can be used to predict that cyclic π systems which have rings containing odd numbers of atoms and which can be dissected into an even AH, will be no more stable than the even AH. The reason is, of course, that formation of a ring containing an odd number of atoms necessitates joining together two atoms of the same set. Since the bond order between these two atoms is zero, first order perturbation theory predicts no change in π energy. An example has already been given in the comparison of azulene and [10]annulene, and several more are presented in Figure 3.24. Actually, many of these molecules are less stable and more reactive than the even AH's from which they may formally be derived. For instance, cyclooctatetraene is a nonplanar molecule; hence, it does not possess the inherent instability and reactivity expected of a planar $4n$ electron system. However, in bridging it to form pentalene, the planarity of the π system is enforced; so pentalene may be anticipated to behave like a true antiaromatic system. After many unsuccessful attempts, the molecule has recently been synthesized and is found to possess all the reactivity expected of an [8]annulene. Even those molecules with $4n + 2$ electron peripheries may be expected to be very reactive toward the proper ionic reagents. For instance, attack by an electrophile at one of the α positions in the five-membered ring of the fourth molecule generates a derivative of the benzotropylium ion, which should possess considerable resonance energy.

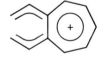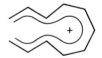

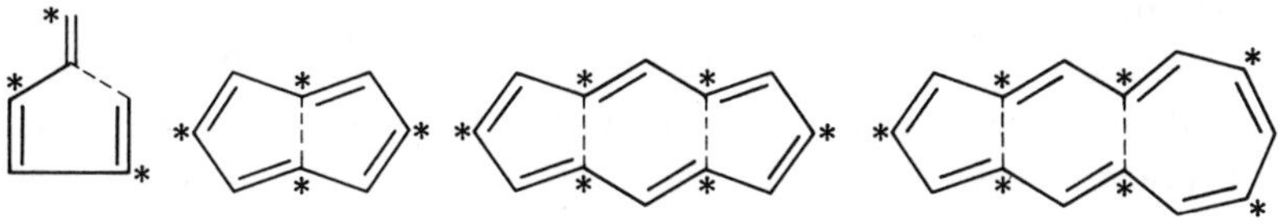

PMO Analysis of Some Molecules That Are Not AH's

Figure 3.24

Problem 3.27.   Calculate the resonance energy, using first order perturbation theory, of benzotropylium relative to that of 1,2-divinyltropylium and undecapentaenylium ions.

Figure 3.25

We have now seen that PMO theory enables us to make many semi-quantitative predictions about π systems with no more work than that involved in using the much less satisfactory resonance theory. However, when using PMO theory in the analysis of a system as two odd AH fragments, we should remember that of the many changes in orbital energies which occur through mixing of MO's in the two fragments, we are ignoring all but one. We justify this by the knowledge that interactions between two bonding or two antibonding orbitals will have little or no effect on the total energy and by the hope that interaction of filled bonding with unfilled antibonding MO's will result in an insignificant energy change compared to that arising from mixing of the NBMO's. Neglect of energy changes due to orbital mixing, other than that between NBMO's, is obviously an approximation; and under certain circumstances it can be a very bad one. For instance, when the NBMO's in two odd AH fragments do not mix, PMO analysis seriously underestimates the stability of the resulting π system. A prime example of this is cyclobutadiene, analyzed as being formed by interaction of allyl with a p orbital (Problem 3.20). Although the p orbital does not mix with ψ_2 of allyl, it does mix with ψ_1 and ψ_3. The interaction of p with ψ_1 lowers the energy of the latter orbital from $\alpha + \sqrt{2}\,\beta$ in allyl to $\alpha + 2\beta$ in cyclobutadiene. In the PMO method, by confining our attention solely to the NBMO's, we ignore this stabilization.

   Another example of how the approximations in the PMO method of analysis can lead to error is found in the comparison of the π energies of benzene and *para*-dehydrobenzene. Three modes of analysis are indicated in Figure 3.26; the first two are in substantial agreement in predicting respectively that the dehydrobenzene is $2 \times 1/\sqrt{3}$ and $2 \times (1/\sqrt{2})^2 \beta$ less stable than benzene. The third mode of analysis predicts that there is no energy difference between the two π systems. The reason that this third mode fails to make the correct prediction is that the choice of NBMO's for use in the analysis has ensured that their energy of interaction across the bridge will be zero. Therefore, whatever change in energy actually results from the bridging is guaranteed to be ignored by this mode of analysis.

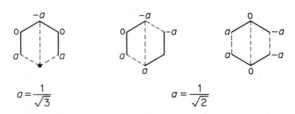

PMO Analysis of the π System of *para*-Dehydrobenzene

Figure 3.26

The energetically unfavorable 1,4 interaction, which the first two modes of analysis correctly predict, comes in the third mode from the out-of-phase combination, $\psi_1 - \psi'_1$, of the bonding allylic MO's, which becomes the degenerate bonding benzene MO with a nodal plane through two bonds. Although one might expect the unfavorable 1,4 interaction in this MO to be balanced by a favorable one in the in-phase combination, $\psi_1 + \psi'_1$, which becomes the lowest benzene MO, this is not quite the case. Mixing of $\psi_1 + \psi'_1$ with the $\psi_3 + \psi'_3$ combination of antibonding MO's results in an increase in the coefficients at the terminal atoms in the in-phase mixture of bonding allylic MO's. Mixing of $\psi_1 - \psi'_1$ with $\psi_3 - \psi'_3$ causes a decrease in those of the out-of-phase bonding combination. These changes are balanced by a decrease in coefficients at the central carbons of the $\psi_1 + \psi'_1$ combination and an increase at these atoms in $\psi_1 - \psi'_1$. Our analysis, which focuses on only what is supposed to be the overwhelmingly largest interaction, the one between the NBMO's, ignores these subtle but important changes.

Problem 3.28.   Use first order perturbation theory on the MO's of benzene to verify the energetic consequences of 1,4 bridging for the π system.

The NBMO's of [4n]Annulenes

The AH theorems can also be used to provide useful information about the NBMO's of [4n]annulenes, which in turn can be applied to the analysis of polycyclic molecules with such a perimeter. A [4n]annulene is an AH, so the pairing theorem is applicable. We know that the lowest MO and its partner, the one highest in energy, are the only two nondegenerate MO's. Excluding these two there are only $(4n - 2)/2 = 2n - 1$ unique energy levels in a [4n]annulene since all the other orbitals come in degenerate pairs. The pairing theorem tells us that corresponding to each level of $E = \alpha + k\beta$, there must be another of energy $E = \alpha - k\beta$. Since $2n - 1$ is an odd number, at least and in this case only one set of degenerate orbitals must be paired with itself and, hence, be nonbonding.

To this nonbonding degenerate pair, (57) and (58) with $E = 0$ again apply. These equations and orthonormality are satisfied by NBMO's of the form

$$\psi_{i,j} = \frac{1}{\sqrt{N}}\left(\sum_{s=1}^{N/2}(-1)^s\phi_s^* \pm \sum_{u=1}^{N/2}(-1)^u\phi_u\right) \tag{60}$$

where N is the total number of atoms in the ring. Alternatively, we can take sums and differences to obtain two NBMO's which are each confined to only one set of atoms

$$\psi'_i = \sqrt{\frac{2}{N}} \sum_{s=1}^{N/2} (-1)^s \phi_s^* \qquad \psi'_j = \sqrt{\frac{2}{N}} \sum_{u=1}^{N/2} (-1)^u \phi_u \qquad (61)$$

In the sense of AH theorem (1), the orbitals of (60) are paired with each other, while in (61) each orbital is paired with itself. As we saw for these two types of NBMO's in cyclobutadiene, which set turns out to be more convenient to employ depends on the problem to be solved.

Although perinaphthyl is an odd AH, so that its NBMO can be found directly, it can also be analyzed by considering it as consisting of a [12]annulene perimeter and a central carbon atom. The set of degenerate [12]annulene NBMO's in which each orbital is paired with itself is most convenient for analyzing this problem, because, as shown in Figure 3.27, only atoms of the unstarred set are attached to the central carbon atom. Therefore, only the NBMO of the perimeter that is confined to this set of atoms interacts with the central carbon. Because the carbon p orbital and the NBMO have the same energy, their interaction lowers their in-phase combination by $3/\sqrt{6}\,\beta$ and raises their out-of-phase combination by the same amount. Since two of the perimeter's 12 electrons originally occupied the degenerate NBMO's, these electrons go into the in-phase combination of the interacting perimeter NBMO and central carbon p orbital. The NBMO of the perimeter that is confined to the starred atoms does not interact with the central atom. Therefore, it remains nonbonding and is occupied only in the perinaphthyl radical and anion.

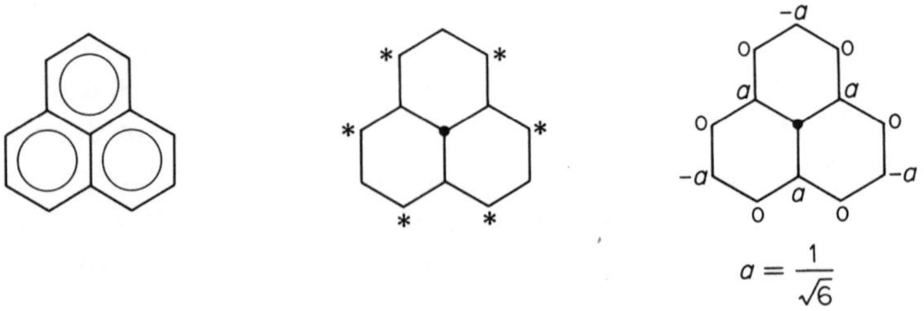

$$a = \frac{1}{\sqrt{6}}$$

PMO Analysis of the Pi System of Perinaphthyl

Figure 3.27

Problem 3.29.   Show that both members of the set of degenerate [12]-annulene NBMO's that are paired with each other interact with the central carbon in perinaphthyl, making them less convenient to use than the set employed above. Can symmetry be used to dictate which set of NBMO's to employ? Verify the results found above by showing, through application of the odd AH theorems directly to perinaphthyl, that the NBMO is indeed confined to the same perimeter atoms as the noninteracting [12]annulene NBMO.

Pyracyclene, in contrast to perinaphthyl, is an even AH. Also, the analysis of pyracyclene requires the [12]annulene NBMO's that are paired with each other; for, as shown below, the self-paired orbitals do not have the proper symmetry, since they lack the planes present in the molecule. In pyracyclene one of the NBMO's of the [4n]annulene periphery interacts with the antibonding MO of the central ethylene, and this results in the lowering of the energy of the NBMO. The other NBMO remains nonbonding, so that pyracyclene is predicted to have an empty NBMO. Indeed, pyracyclene is found to undergo facile two-electron reduction, the potential at which the first electron is accepted being almost identical with that required for the reduction of the triphenylmethyl radical.

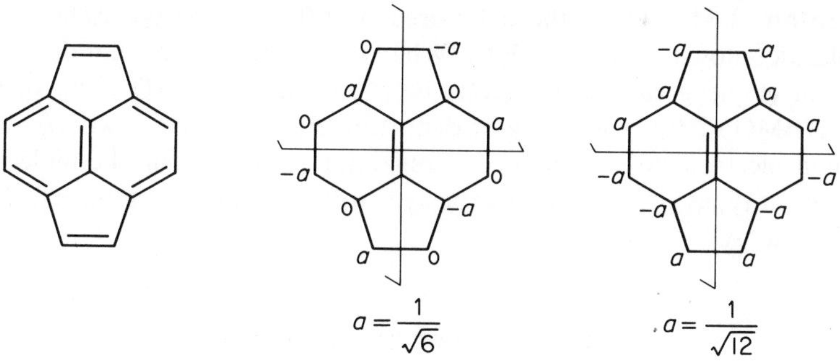

PMO Analysis of the π System of Pyracyclene

Figure 3.28

Problem 3.30.   Which set of perimeter NBMO's is appropriate for the analysis of the [16]annulene analog of pyracyclene? Do you predict that its oxidation or its reduction should be unusually easy?

Although it is generally true that interaction of a [4n]annulene periphery with a central ethylene, as in pyracyclene, occurs with the antibonding MO of the olefin when n is odd, the results with n even differ depending on whether n is divisible by four, as it is in the [16]annulene analog of pyracyclene (Problem 3.30). If n is not divisible by four, as in the hypothetical molecule shown in Figure 3.29, then although there exists an NBMO of the [4n]annulene perimeter with symmetry appropriate for interaction with the ethylene bonding MO, such interaction is expected to be small, since the NBMO has its nodes at the carbons that are adjacent to those of the ethylene.

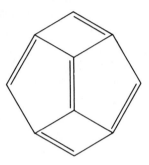

Figure 3.29

Problem 3.31.   Draw the self-paired NBMO's of the perimeter in the molecule shown in Figure 3.29. What distortions in the bonds to the central ethylene can cause mixing between one of its MO's and one of the NBMO's? For each type of distortion state whether it will result in a high electron affinity or a low ionization potential for the molecule. Are both types of distortions expected to have about the same effect on the π energy?

Some Further Applications
of the Theorems for Odd AH's

Thus far, we have used the AH theorems solely for calculating energies via the PMO method. Another use to which the NBMO coefficients can be put is in calculating charge densities for odd AH's. Since all atoms are neutral in a neutral AH, the squares of the coefficients in the NBMO give the net negative charge produced at each atom by adding an electron to the π system to form an anion. The density of the electron hole left at each atom when the NBMO is emptied to form a cation can similarly be computed, and is equal to the net positive charge at each atom. For instance, suppose we wish to know how to best stabilize a benzyl cation. Reference to the coefficients in the NBMO (Figure 3.18) shows that removal of the electron from this orbital in the benzyl radical leaves a net positive charge, or electron "hole," of 1/7 at the *para* and each of the two *ortho* positions and 4/7 at the benzylic carbon. First order perturbation theory tells us then that the best place for an electron donating substituent is at the benzylic carbon; such a substituent will be somewhat less effective at the *ortho* or *para* positions, and it will be completely ineffective at the *meta* positions. The same is, of course, true for an electron withdrawing substituent in the anion, since the anion has the identical distribution of negative charge.

Problem 3.32.   Considering the relative electronegativities of nitrogen and carbon, which of the picolines (monomethyl pyridines) do you expect to be most acidic?

Problem 3.33.   Demonstrate that in the 9-substituted anthracene shown in Figure 3.30 a stabilizing substituent attached at the 10 position would be as effective as one attached to the methylene carbon in both the anion and cation.

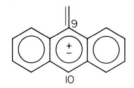

Figure 3.30

The NBMO coefficients can also be used to help predict the preferred position of electrophilic attack on aromatic systems. For instance, if the π system of α-naphthol can be treated crudely as that of an α-naphthylcarbinyl carbanion, and assuming that electrophilic attack occurs at the carbon of highest π electron density, we predict, on the basis of the NBMO coefficients calculated for this system (Figure 3.18), that attack at the 2 and 4 positions should be about equally facile. In β-naphthol attack is predicted from the results of Problem 3.21 to occur largely at the 1 position.

Problem 3.34. Consider the π system remaining after electrophilic attack on naphthalene at the 2 and at the 4 positions. In which case could an electron donating substituent at the 1 position, as in α-naphthol, best stabilize a positive charge?

Of course, another important factor in assessing where electrophilic substitution will occur – especially if the intermediate, where one carbon is removed from the π system, is formed irreversibly – is the relative energies of the possible intermediates.[22] However, we have no way of directly calculating the stabilities of the odd AH fragments that are left after removing one carbon from conjugation in an even AH. Nevertheless, we can calculate the energy liberated on restoring conjugation by allowing each odd AH fragment to interact once again with the carbon atom that was removed. This energy is just the negative of that required to remove the atom from conjugation. If various atoms are removed from conjugation in the same molecule, the differences in the energy required to remove different atoms must be due to the differences in energy of the remaining odd AH fragments. Moreover, since electrons in NBMO's do not influence relative stabilities of odd AH fragments, it does not matter whether the atom is removed from the π system by electrophilic, radical, or nucleophilic attack.

A calculation, using this technique for estimating the energy necessary to remove an atom from conjugation, is illustrated in Figure 3.31 for naphthalene. It is predicted that, since removal of the 1 position from conjugation causes an increase in π energy of $(-2 \times 3/\sqrt{11})\beta$ while that

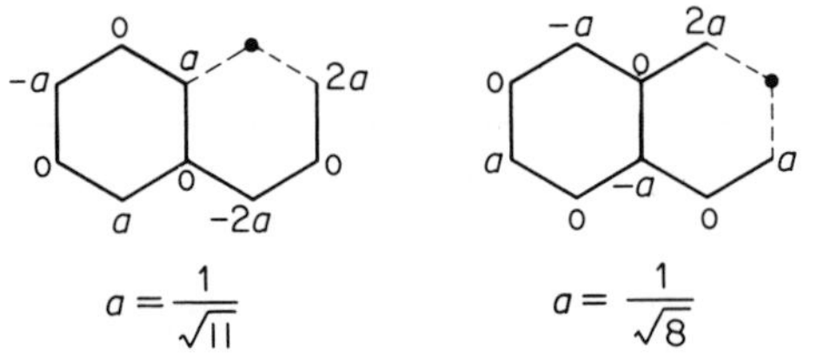

PMO Analysis of the Relative Energies of Two Odd AH's

Figure 3.31

[22] The rates of formation of the intermediates will, needless to say, depend not on their energies but on the energies of the transition states leading to them. Nevertheless, since the transition states are expected to resemble the intermediates, we can use the relative energies of the latter to approximate those of the former.

of the 2 position causes an increase of $(-2 \times 3/\sqrt{8})\beta$, attack at the 1 position should be favored. Our predictions regarding the position of electrophilic attack on naphthalene and the naphthols find experimental confirmation in the chemistry of these compounds.

Problem 3.35.   Predict the relative rates of protonation at the three types of carbons in anthracene. Do you expect anthracene to be more easily protonated than naphthalene?

Relationship Between Resonance and Molecular Orbital Theory

We have seen that PMO theory enables one to make quite accurate predictions, based on the principles of perturbation theory and the properties of AH's, particularly odd AH's. Many of these predictions can also be made using resonance theory, when cognizance is taken of the Hückel $4n + 2$ rule. There is an excellent reason for this correspondence; for it is possible to prove that there is a theoretical relationship between the coefficients in the NBMO of an odd AH and the number of resonance structures that can be written for an even system formed from it by attachment of an additional carbon atom.[23] Providing that an odd AH does not contain a ring of $4n$ atoms, the unnormalized NBMO coefficient at each atom is proportional to the number of resonance structures that can be drawn for the molecule formed by attaching to that atom a carbon bearing a p orbital. If the added atom is attached to the odd AH at more than one point, the number of resonance structures in the molecule formed is proportional to the sum of the absolute values of the coefficients at the points of attachment.

Problem 3.36.   Use the above theorem to verify that there is only one resonance structure that can be drawn for decapentaene, two for cyclodecapentaene and azulene, and three for naphthalene and bicyclo[6.2.0]-decapentaene. Show that for this last molecule the absolute value of the sum of the coefficients is a better indicator of the number of resonance structures that actually contribute to its stability than the sum of the absolute values.

[23] See M. J. S. DEWAR and H. C. LONGUET-HIGGINS, *Proc. Roy. Soc.* (London) **A214**, 482 (1952) for the proof of this theorem.

Problem 3.37. Resonance theory can be used to estimate the relative double bond character and, consequently, the relative bond lengths in conjugated systems which do not contain a $4n$ membered ring. For instance, in naphthalene (Figure 3.10) two of the three resonance structures place a double bond between α and β carbons; all other carbon atoms are connected by a double bond in only one resonance structure. Experimentally, it is found that the bonds in naphthalene are roughly 1.42 Å long, except for those between α and β atoms, which are 1.37 Å. Use the theorem regarding the relationship between the coefficients in the NBMO of an odd AH and the number of resonance structures that can be written for an even AH formed from it by adding a carbon atom to show theoretically why resonance theory can be used to estimate the relative double bond character in such molecules. [Hint: Equate the PMO expression for the energy of the π bonds formed on attaching a carbon atom to an odd AH at several points with the correct expression, in terms of the bond orders, for the energy of these π bonds in the molecule formed. Does such a relationship hold exactly? Do you therefore expect a *quantitative* relationship to exist between fractional double bond character, as estimated by counting resonance structures, and bond length? Show how PMO theory can be used to predict the relative bond lengths in naphthalene.]

Problem 3.38. Another relationship exists between resonance theory and the NBMO coefficients in an odd AH. The sum of the absolute values of the coefficients is proportional to the number of resonance structures that can be written for the odd AH, excluding those which place $4n$ electrons around a ring. Use this theorem to confirm that seven resonance structures may be drawn for the species formed by removing an α atom from naphthalene but that only six can be found for the fragment resulting from removal of a β atom. How many resonance structures can be drawn for perinaphthyl? How many are predicted from the coefficients in its NBMO? If there is a discrepancy, explain.

APPENDIX FOR CHAPTER 3

Theorems for Neutral AH's

Since we can write any MO as an LCAO

$$\psi_i = \sum_r c_{ir}\phi_r \qquad (62)$$
$$\text{(all atoms)}$$

a well known theorem of algebra tells us that we can then express any AO as a linear combination of MO's

$$\phi_s = \sum_j c_{sj}\psi_j \qquad (63)$$
$$\text{(all MO's)}$$

If the AO's as well as the MO's are orthonormal, it is easy to show that the coefficient c_{is} in (62) is equal to c_{si} in (63). Multiplying (62) by ϕ_s and (63) by ψ_i, integrating both equations, and equating the two expressions for $\int \phi_s \psi_i \, d\tau$ demonstrates that $c_{is} = c_{si}$. Making this substitution in the expressions for ϕ_r and ϕ_s from (63) and using the assumption of orthonormality for the AO's allows us to write

$$\int \phi_r \phi_s \, d\tau = \delta_{rs} = \sum_i \sum_j c_{ir}c_{js} \int \psi_i \psi_j \, d\tau = \sum_i \sum_j c_{ir}c_{js} \delta_{ij} = \sum_i c_{ir}c_{is} \qquad (64)$$
$$\text{(all MO's)} \qquad\qquad \text{(all MO's)} \qquad\qquad \text{(all MO's)}$$

where $\delta_{rs} = 1$ for $r = s$ and $\delta_{rs} = 0$ for $r \neq s$. Dividing the sum over all the MO's in (64) into sums over the bonding, nonbonding, and antibonding MO's, we have for $r = s$

$$\sum_i c_{ir}^2 + \sum_{i'} c_{i'r}^2 + \sum_{i''} c_{i''r}^2 = 1 \qquad (65)$$
$$\text{(bonding)} \quad \text{(nonbonding)} \quad \text{(antibonding)}$$

but using the pairing theorem $c_{i'r}^2 = c_{ir}^2$ or $(-c_{ir})^2$, so (65) becomes

$$2 \sum_i c_{ir}^2 + \sum_{i'} c_{i'r}^2 = 1 \qquad (66)$$

(bonding) (nonbonding)

Since in a neutral AH there are two electrons in every bonding MO, one in each nonbonding MO, and none in the antibonding orbitals, (66) can be rewritten

$$\sum_i n_i c_{ir}^2 = q_r = 1 \qquad (67)$$

(all MO's)

which proves that the electron density on any atom in a neutral AH is unity.

Problem 3.39.   Use (64) to prove the theorem that the bond order between two atoms of the same set in a neutral AH vanishes.

FURTHER READING

A few of the standard texts on the application of MO theory to the π systems of organic molecules are: J. D. ROBERTS, *Notes on MO Calculations*, Benjamin, 1962; A. STREITWEISER, *Molecular Orbital Theory for Organic Chemists*, Wiley, New York, 1961; L. SALEM, *Molecular Orbital Theory of Conjugated Systems*, Benjamin, New York, 1966, and MICHAEL J. S. DEWAR, *The Molecular Orbital Theory of Organic Chemistry*, McGraw-Hill, New York, 1969. Chapter 6 of Dewar's book gives a comprehensive treatment of the perturbational MO (PMO) method of treating π systems, for whose development Dewar, along with Coulson and Longuet-Higgins, deserves credit.

The original paper on bond alternation in $4n + 2$ annulenes by H. C. LONGUET-HIGGINS and L. SALEM, *Proc. Roy. Soc.* (London) **A251**, 172 (1959), also contains calculations on the extent of and energy gain from alternation and demonstrates that very long polyenes and polyene radicals should also show bond alternation.

Several relationships between resonance and MO theory are discussed in W. C. HERNDON, *Tetrahedron*, **29**, 3 (1973) and references therein.

4

The
Woodward–Hoffmann Rules
for Pericyclic Reactions

Pericyclic reactions are those reactions in which a closed loop of orbitals participate in the transition state. Examples of these are the cyclo-addition of two ethylenes to form cyclobutane, the Diels-Alder reaction, a 1,5-hydrogen shift in a conjugated diene, and the ring opening of cyclo-hexadiene to hexatriene. Not all of these reactions actually occur. For instance, in contrast to the Diels-Alder reaction, the cycloaddition of two ethylenes with simultaneous formation of both new bonds is unknown. Ethylene dimerization in which the bonds are made in a stepwise fashion, leading to formation of a discrete biradical intermediate, is known, particularly with fluorinated ethylenes. However, this process will not concern us here. In this chapter we will only be concerned with pericyclic reactions, and a reaction in which bonds are formed stepwise does not belong to this class.

Not only do some of the above pericyclic reactions fail to take place, but others occur in only one of two a priori possible ways. For instance, cyclohexadiene in opening could conceivably rotate both methylene carbons in the same direction (conrotatory mode) or in opposite directions (disrotatory mode). In fact, only disrotatory opening is observed in thermal reactions. This does not appear to be a steric effect, for the photochemically initiated opening is conrotatory. In addition, the thermal opening of cyclobutene is conrotatory.

In this chapter we shall examine the theoretical analysis, the development of which is primarily due to Woodward and Hoffmann, of these pericyclic reactions. Since these reactions involve the formation of sigma bonds, the reader may question why they are treated in the middle of our discussion of π electron theory and not in some later chapter. The reason is that the concept of aromaticity plays an important part in the generalized theoretical treatment of these reactions. In fact, Dewar has suggested that the simplest statement of the Woodward-Hoffmann rules is that concerted pericyclic reactions take place via "aromatic transition states." Why this is the case and what constitutes such a transition state is the subject of this chapter.

Cycloaddition Reactions

Consider what happens as two ethylene molecules are brought together in the proper geometry for cyclobutane formation. While the sigma overlap between the two molecules increases, the π overlap between the AO's within each molecule decreases as the double bonds become the single bonds of the four membered ring. Consequently, the magnitude of β_σ for the sigma bonds which are forming increases, until, at the end of the reaction, it is greater than that of β_π for the remaining π interactions. Thus, at some point along the reaction coordinate the β's for π and sigma interactions must be equal. Therefore, at this point along the reaction coordinate, the MO's for the system must have the same *pattern* of energy levels as cyclobutadiene.[1] Intuitively, we might then expect that whatever destabilizes cyclobutadiene will also destabilize the transition state for the dimerization of two ethylenes.

Problem 4.1. Can you make a similar argument to explain why the Diels-Alder reaction does occur?

---

[1] The actual energies will be different, since the resonance integrals in the two systems will be different.

We can get some idea of what causes the destabilization of the transition state in the dimerization of two ethylenes from the application of perturbation theory to this problem. We have really already done this in the previous chapter when we considered the π interaction of two ethylenes. The symmetry considerations we used then to establish that the bonding orbital in one ethylene would only interact with the bonding orbital in the other, remain unchanged when we bring two ethylenes together so they overlap in a sigma fashion. The chief difference is that the magnitude of β will be less in the transition state than β_π in the reactants, as we have just discussed. Therefore, the previous diagram we drew should be modified by lowering the bonding levels and raising the antibonding levels in the reactants, relative to the MO's produced by their mixing. The resulting orbital interaction diagram is drawn in Figure 4.1.

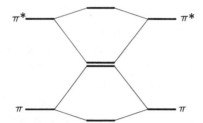

Orbital Interaction Diagram for Ethylene Dimerization

Figure 4.1

From this diagram we can gain some insight into why the cyclobutadienoid pattern of energy levels present in the transition state is so unstable. We see that it is achieved by progressively raising a filled orbital in energy. True, another filled orbital is somewhat lowered in energy; but as we can see, this does not offset the large increase in energy of the first. The large decrease in energy of the empty orbital will, of course, not affect the net energy of the system, which is expected to rise steeply to the transition state.

Since we know from symmetry that the mixing of the ethylene MO's will be equal and result in an in-phase and an out-of-phase combination of each, we can construct the resultant MO's, whose changes in energy are plotted schematically in Figure 4.1. These are drawn in Figure 4.2. We can see that the offending MO which causes the net energy rise is the $\pi_1 - \pi_2$ combination of bonding MO's, which, as the reaction progresses,

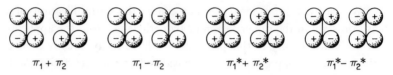

$$\pi_1 + \pi_2 \qquad \pi_1 - \pi_2 \qquad \pi_1{}^* + \pi_2{}^* \qquad \pi_1{}^* - \pi_2{}^*$$

Symmetry Combinations of π MO's in Ethylene Dimerization

Figure 4.2

involves a progressively serious antibonding sigma interaction. In fact, it appears that this orbital must eventually become one of the antibonding sigma MO's in the product. By contrast, the in-phase combination of antibonding MO's is greatly stabilized because of an increasingly bonding sigma interaction, and it seems destined to eventually become one of the bonding MO's in cyclobutane. However, as we have already noted, this orbital is initially empty and so does not affect the total electronic energy, unless it subsequently becomes filled.

We can verify how the orbitals of the reactant are transformed into the orbitals of the product by construction of an orbital correlation diagram. Two planes of symmetry m_1 and m_2 are maintained throughout the reaction,[2] and we begin by classifying the symmetry of each orbital in reactant and product with respect to these planes. We then correlate orbitals of the same symmetry in reactant and product, using the rule that those of the same symmetry will not cross (since they will mix to prevent this), if the same symmetry type occurs more than once. The results for the dimerization of ethylene are shown in Figure 4.3.

The correlation diagram confirms our suspicion about the ultimate fate of $\pi_1 - \pi_2$; it does become an antibonding orbital in the product, and $\pi_1{}^* + \pi_2{}^*$ becomes a bonding MO. Thus, we gain a little more insight into the significance of the cyclobutadienoid pattern of energy levels in the transition state—it arises because a filled orbital in the reactant is correlated with an empty one in the product and vice versa. When ethylene tries to dimerize, a filled bonding orbital is transformed into an antibonding cyclobutane MO. As the reaction proceeds, this unfavorable correlation becomes more and more costly energetically so long as this MO remains filled. However, if one electron were removed from $\pi_1 - \pi_2$

---

[2] A third symmetry plane, that which contains all the carbon atoms, is also maintained throughout the reaction. However, no bonds are made or broken across it, and classification of the orbitals of reactant and product with respect to a symmetry element across which no bonds are made or broken provides no information as to whether the MO's of the reactant are transformed smoothly into those of the product.

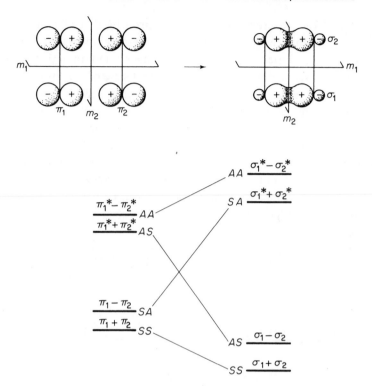

Orbital Correlation Diagram for Formation of Cyclobutane by Ethylene Dimerization

Figure 4.3

and placed in $\pi_1^* + \pi_2^*$, the correlation diagram predicts that the increase in energy of the former MO would be balanced by the decrease in that of the latter; and the overall reaction should be thermally neutral. In fact, the *photochemical* formation of a cyclobutane ring from two ethylenic units is a well known reaction.

What exactly happens at the point at which the two orbitals cross is an interesting question. From the orbital correlation diagram one would expect that the dimerization of two ethylenes might lead to a doubly excited product unless the electrons "jump" into the other orbital at the crossing point. In order to investigate this possibility, we must construct a correlation diagram for the electronic states, formed by assigning electrons to the various orbitals. This time we classify the *states* according to their symmetries, which we can determine from the symmetries of the occupied orbitals in them. This is accomplished by multiplying the symmetry of the wavefunction of each electron by that of every other, using the rules that the product of two symmetric or two antisymmetric functions is symmetric, while that of a symmetric and an antisymmetric function is antisymmetric.[3] Thus, in the ground state of two ethylenes the symmetry with respect to plane m_1 is $S \times S \times S \times S = S$, while that with respect to plane m_2 is $S \times S \times A \times A = S$; so the state $(\pi_1 + \pi_2)^2(\pi_1 - \pi_2)^2$ has SS symmetry. The first excited state, $(\pi_1 + \pi_2)^2(\pi_1 - \pi_2)(\pi_1^* + \pi_2^*)$, has $S \times S \times S \times A = A$ symmetry to m_1 and $S \times S \times A \times S = A$ symmetry to m_2; we designate it AA. The symmetry of the doubly excited $(\pi_1 + \pi_2)^2(\pi_1^* + \pi_2^*)^2$ is SS. The symmetries of the important states in the product can similarly be determined and are shown in Figure 4.4.

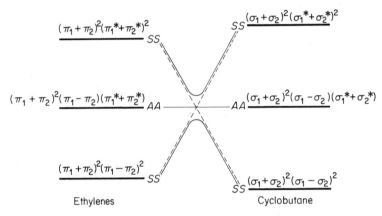

State Correlation Diagram for Cyclobutane Formation by Ethylene Dimerization

Figure 4.4

From the correlation diagram we see that although the orbitals in the ground state of the reactant are correlated with those of a doubly excited state of the product and vice versa, as indicated by the dotted lines, the two wavefunctions are prevented from actually correlating in this fashion by the fact that they have the same symmetry and consequently will not cross. Thus, the solid line represents the actual behavior of the ground state in the reactant, which does, in fact, correlate with the ground state in the product. Nevertheless, the *orbital* correlation with a doubly excited state of the product does have a profound effect on the reaction,

[3] In Chapter 6 we will see the origin of these rules.

for it results in the large energy maximum near the point at which the states would have crossed, were it not for the noncrossing rule.

This rule is applied here on the assumption that the Hamiltonian contains some term which will always allow mixing between functions of the same symmetry, lowering the energy of the lower and raising that of the upper, so that crossing is averted. What is this term that apparently allows unfavorable correlation of one-electron wavefunctions (orbitals) but prevents the crossing of many-electron wavefunctions (states)? It is, in fact, none other than the electron–electron repulsion operator that causes the mixing of the electronic configurations. Since we have not taken explicit account of this term in the Hamiltonian, we cannot yet derive how or why it causes this mixing; however, let us foreshadow what we will find in Chapter 7. When we explicitly include the electron repulsion operator in the Hamiltonian, we shall find that mixing of many-electron wavefunctions, or as it is more commonly called, configuration interaction, occurs because it results in a decrease in electron repulsion. We saw in Chapter 1 that the form of the LCAO-MO wavefunction for the ground state of the hydrogen molecule results in an equal mixture of ionic and covalent terms. We shall show in Chapter 7 that mixing into the ground state wavefunction the proper amount of the wavefunction for the appropriate doubly excited configuration causes a decrease in the ionic character and hence the energy of the ground state wavefunction. A situation in which configuration interaction is of crucial importance is in the proper description of a singlet diradical. Here two electrons must be placed in degenerate NBMO's. Since the orbitals occupied by these two electrons are nonbonding, any ionic terms in the wavefunction for these two electrons only serve to increase the energy. The proper description of this situation in MO theory requires the equal mixing of what are, ordinarily, the ground and appropriate doubly excited configurations. Since in a diradical these electronic configurations have the same energy, they can mix equally; and the mixing removes all the ionic terms from one of the two linear combinations.

We note that at the point of avoided crossing in the state correlation diagram, Figure 4.4, equal mixing must occur. It occurs because $(\pi_1 - \pi_2)$ at this point is a nonbonding orbital, since all its bonding π character is cancelled by antibonding sigma interactions. $(\pi_1^* + \pi_2^*)$ is also nonbonding, its antibonding π being cancelled by its bonding sigma interactions. Thus, the two configurations of the same symmetry involving these two orbitals must have the same energy at the point where their crossing is avoided and therefore mix equally to give the correct description of what must be, at the point of avoided crossing—a diradical! The concerted dimerization of two ethylenes (or fragmentation of cyclobutane) must, thus, proceed through a transition state which has the electronic structure and energy of a diradical.[4]

The high energy of the potential curve for the ground state near the point of avoided crossing not only creates a large barrier to the reaction but also brings the ground state curve close to that for the first excited state of the system. The crossing of these two curves is, of course, allowed, since they have different symmetries; in fact, it almost certainly occurs since highly excited vibrational levels of the ground state will lie at the same energy as some low levels of the excited state. The result is that a molecule on the excited state curve may undergo a nonradiative transition to the ground state, and a molecule on the ground state curve may similarly cross to the lowest excited state of the system. There is experimental evidence for the occurrence of both these events. Perhaps the most striking example comes in the thermal decomposition of dioxetanes, a highly exothermic reaction which produces one molecule of the carbonyl fragmentation product in an excited state. Since this same excited state can be populated by photochemical excitation of the carbonyl compound, the thermal reaction which produces it has been appropriately called "photochemistry without light" by Emil White.[5] A similar type of fragmenta-

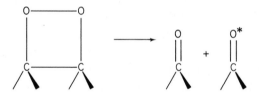

Figure 4.5

---

[4] This diradical is not tetramethylene—the diradical that results from forming one sigma bond fully and leaving the remaining two electrons in noninteracting AO's. The diradical at the point of avoided crossing still maintains both planes of symmetry and can be shown to have its nonbonding electrons, one each, in MO's analogous to the NBMO's in cyclobutadiene with nodes through atoms diagonally across the ring. Of course, a reaction path which passes through a diradical of lower energy may exist and consequently be the one the reaction preferentially follows.

[5] For a review of chemically produced excited states see *Angew. Chem. Int. Edit.*, **13**, 229 (1974).

tion reaction appears to be responsible for the familiar chemiluminescence produced by fireflies.

We have now examined in some detail a reaction which has a cyclobutadienoid array of orbitals in the transition state, seen why such an array is energetically unfavorable, and explored some of its consequences. Now we turn to the Diels-Alder reaction, which we may anticipate will possess a benzenoid array of orbitals. Use of second order perturbation theory allows us to construct a diagram (Figure 4.6) for the orbital interactions in the reaction of ethylene with butadiene.

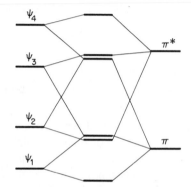

Orbital Interaction Diagram for the Diels–Alder Reaction
of Ethylene with Butadiene

Figure 4.6

We have, of course, obtained a very similar diagram in the last chapter when we considered the interaction of a butadiene with an ethylene in a π, rather than in a σ fashion as we do now. As in the case of two ethylene molecules, the lowest bonding MO's, both of which have the same symmetry, mix to give two combinations, one of which is strongly bonding and the other of which is antibonding in the region where the MO's of the reactants interact. However, the rise in energy of the out-of-phase combination, $\pi - \psi_1$, is mitigated by further mixing with the empty ψ_3, which removes the antibonding interaction as shown in Figure 4.7. Moreover, throughout the reaction, ψ_2 is stabilized by mixing with π^*, giving an orbital which is strongly bonding between the reactants. Unlike the dimerization of two ethylenes, the Diels-Alder reaction can proceed because of the availability of *low-lying unfilled* orbitals in each reactant of the *appropriate symmetry* for interaction with *high-lying filled* levels in the other.

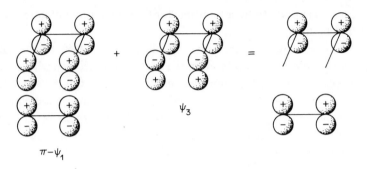

Mixing of π with ψ_1 and ψ_3 in the Diels–Alder Reaction

Figure 4.7

Problem 4.2. (a)   Explain why one might expect extraordinary reactivity from cyclobutadiene acting as the diene component in a Diels-Alder reaction.

(b)   If butadiene is substituted with an electron donating group (i.e. one containing a filled orbital of relatively high energy) and ethylene is substituted with an electron withdrawing group (i.e. one containing a low-lying unfilled orbital), between which orbital of the butadiene and which of the ethylene will the principal interaction occur in a Diels-Alder reaction? Predict the relative rate of the reaction between such substituted molecules compared to that between ethylene and butadiene.

(c)   The relative size of the coefficients in ψ_2 and ψ_3 for butadienes substituted at C-1 and C-2 with an electron donor (D) is shown schematically in Figure 4.8, where L represents a large coefficient and s a

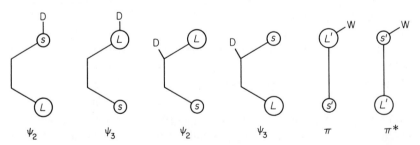

Relative Size of Coefficients in ψ_2 and ψ_3 in Dienes Substituted with an Electron Donating Group and in π and π^* in an Ethylene Substituted with an Electron Withdrawing Group

Figure 4.8

small one. Also shown is the relative size of the coefficients in an ethylene substituted with an electron withdrawing group (W). Show that in a Diels-Alder reaction between the 1-substituted butadiene and the ethylene a regioselectivity favoring a 3,4 disposition of the substituents in the cyclohexene product will be observed, while with the 2-substituted diene, a 1,4 orientation will predominate. [Hint: Apply perturbation theory. You will need to demonstrate mathematically that $LL' + ss' > Ls' + L's$.]

Problem 4.3.   Use the plane of symmetry which bisects each of the reactants to classify according to symmetry the orbitals in the reactants and products in the Diels-Alder reaction. Show by constructing a correlation diagram that ψ_1 and ψ_2 become the symmetric and antisymmetric combination of sigma bonding MO's respectively. Show that π in the ethylene reactant correlates with π in the cyclohexene product, so that all the bonding MO's of the reactant are correlated with bonding MO's of the product. What do you predict about the feasibility of a photochemical Diels-Alder reaction if the excitation removes an electron from ψ_2 and places it in ψ_3?

The correlation diagram for the Diels-Alder reaction is quite different from that for ethylene dimerization. In the cycloaddition reaction involving four π electrons a bonding MO in the reactant correlates with an antibonding MO in the product, while in the six electron reaction, bonding MO's correlate only with bonding MO's. Just as we saw previously that a cyclobutadienoid pattern of energy levels along the reaction path implied the existence of a bonding-antibonding orbital correlation between reactants and product, now we see that a benzenoid pattern is associated with correlation of only bonding orbitals in reactants and product, as shown schematically in Figure 4.9.

In general, cycloaddition reactions may be expected to proceed smoothly when $4n + 2$ electrons are involved in the reaction. In such an "aromatic" transition state only filled levels of reactants are transformed into filled levels of the product, so no large barrier to the reaction is expected. In contrast, $4n$ electron cycloaddition reactions may be expected to have transition states that are essentially diradical in nature, since they involve the correlation of a bonding orbital in the reactant with an antibonding level in the product. Thus, these latter reactions are expected to require larger activation energies, because when the "antiaromatic" transition state is reached, a bonding orbital in the reactants, containing two electrons, has been transformed into a nonbonding MO.

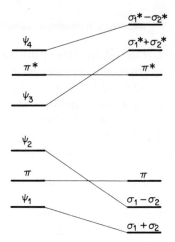

Orbital Correlation Diagram for the Diels–Alder Reaction

Figure 4.9

Problem 4.4. Show that in the dimerization of two butadienes to cyclooctadiene, each reactant molecule possesses unfilled orbitals of the requisite symmetry to allow mixing with filled levels in the other reactant. Construct a correlation diagram to show that the reaction is nevertheless expected to require a large activation energy for concerted bond formation. Can an orbital interaction diagram be used to make this same unequivocal prediction?

Sigmatropic Reactions

A sigmatropic change of order $[i, j]$ is defined by Woodward and Hoffmann as "the migration of a σ bond, flanked by one or more π electron systems, to a new position whose termini are $i - 1$ and $j - 1$ atoms removed from the original bonded loci, in an uncatalyzed intramolecular process." Examples are the already alluded to 1,5 hydrogen migration and the Cope and Claisen rearrangements. The first in the notation of Woodward and Hoffmann is a [1,5] sigmatropic reaction, and the second and third are of order [3,3]. Since only a very small number of sigmatropic reactions will preserve a symmetry element throughout the reaction, correlation diagrams do not provide a very useful way of analyzing this class of pericyclic reactions. Noting that a sigmatropic reaction involves the transfer of an atom or group from one end of an odd AH to the other, a quite natural way to proceed is to use our PMO method of analysis of the previous chapter to estimate the relative stability of the transition state in terms of interactions between NBMO's of odd AH's.

For example, the analysis of a 1,5 hydrogen shift is almost identical to the analysis of benzene in terms of the interaction of a carbon atom and pentadienyl. The only difference is that the $2p$ orbital of the carbon atom is replaced by a $1s$ hydrogen orbital, which is partially bonded to only the lower lobes of the p orbitals of the terminal carbons. Woodward and Hoffmann term this particular geometry "suprafacial" to distinguish it from another possibility, in which the hydrogen interacts with a top lobe at one end and a bottom lobe at the other, for which they reserve the name "antarafacial."

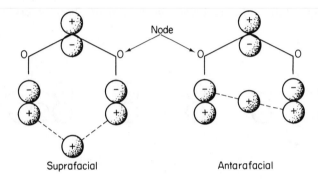

Suprafacial   Antarafacial

PMO Analysis of a 1,5 Hydrogen Shift in 1,3-Pentadiene

Figure 4.10

The two possible geometries for a 1,5 hydrogen shift are shown above along with the phases in the pentadienyl NBMO. In contrast to the suprafacial reaction, which can proceed smoothly with bonding throughout the transfer of the hydrogen atom, the antarafacial reaction, in addition to its geometric problems in maintaining adequate overlap between orbitals, attempts the transfer between lobes of unlike sign in the pentadienyl NBMO. Therefore, midway through the latter reaction the bonding and antibonding interactions between the pentadienyl NBMO and hydrogen atom approximately cancel, and the transition state must closely resemble a diradical.[6]

---

[6] In the sigmatropic transfer of an atom, even when there is no net interaction in the transition state between the NBMO and the AO of the atom, the latter can still interact with other π MO's of appropriate symmetry. Interactions between the atom being transferred and such orbitals, "subjacent" to the NBMO, should in principle, therefore, slightly favor concerted over noninteracting biradical transition states for sigmatropic reactions, even when an atom is transferred between AO lobes which have opposite sign in the NBMO.

This does not mean that all antarafacial reactions necessarily require high energy. Consider the allylic system shown in Figure 4.11. Although geometrically unfavorable, an antarafacial transfer of a hydrogen atom could occur between two lobes of the same sign, which is not the case for the suprafacial reaction. However, a suprafacial reaction could occur with bonding interactions in the transition state between both termini of the allylic system and the atom being transferred, if the migrating group possessed a p orbital, as shown. Thus, a [1,3] sigmatropic shift, if it occurs suprafacially, should result in inversion of configuration in a migrating carbon. Such a stereochemical event attending suprafacial [1,3] sigmatropic shifts in thermal reactions has, in fact, been observed experimentally.

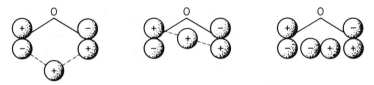

PMO Analysis of 1,3 Sigmatropic Reactions

Figure 4.11

Problem 4.5. Although an antarafacial [1,3] sigmatropic reaction is probably impossible for geometric reasons, an antarafacial reaction in longer chains might be feasible. Account for the fact that although 1,3 hydrogen shifts are unknown in thermal reactions, 1,5 shifts are common and 1,7 shifts have also been observed. Show that the Cope and Claisen rearrangements ought to proceed readily, provided that they occur suprafacially on both components. Would an antara-antara Cope rearrangement be difficult electronically? geometrically?

Problem 4.6. Account for the fact that [1,2] sigmatropic shifts are common in carbocations, but not in carbanions.

In our discussion of sigmatropic shifts the reader may have noted that we have encountered some apparent anomalies. For instance, we have been led to believe that a reaction which involves a six electron transition state should be facile, since the fact that it is an "aromatic" transition state implies that bonding orbitals in the reactants are smoothly transformed into bonding orbitals in the product. Indeed, a suprafacial 1,5 hydrogen shift is predicted by our perturbational analysis to be a facile reaction; but another six electron transition state, that for the antarafacial reaction, is predicted to be diradical in nature. Similarly, despite the fact that the 1,3 shift of an atom requires a four electron transition state, we were able to find two ways of accomplishing this transfer which allowed continuous bonding interactions in the transition state. Only the suprafacial reaction that retained the stereochemistry at the migrating group appeared to pass through a diradical-like transition state.

These anomalies occur only when an antarafacial component[7] is present, since a p orbital that migrates with inversion also is really being used antarafacially.[8] Systems which involve antarafacial interactions[7] must follow a different set of rules than those which do not involve such interactions; in fact, the rules must be the exact reverse of those followed by the Hückel systems with which we have heretofore dealt exclusively. Transition states containing an odd number of antarafacial interactions apparently allow smooth transformation of orbitals of the reactant into those of the product when $4n$ electrons are involved but are diradical in nature when $4n + 2$ electrons participate. Why this is the case is revealed in the correlation diagrams for electrocyclic reactions.

Electrocyclic Reactions

We have already alluded to the empirical fact that cyclobutene ring opening takes place in a conrotatory mode. Since the transition state for the ring opening contains four electrons, we anticipate that the conrotatory mode must involve an antarafacial component, or the reaction would not proceed; for a Hückel system containing $4n$ electrons would imply a diradical transition state of high energy. The transition state for the conrotatory opening is shown in Figure 4.12, and we confirm the fact that it does indeed necessitate the use of one antarafacial component. For instance, if as in **A** we wish to use the π bond suprafacially, we must use one of the small sigma back lobes and one of the large front lobes; thus the sigma bond is an antarafacial component. We might also

[7] More precisely, an odd number of them; (See Problem 4.5.).

[8] We may generally define an antarafacial component as one in which AO lobes of opposite phase are utilized in a reaction. The phases of the AO's of a reaction component, which is not an atom, are, for the purposes of the above definition, assigned so that the maximum number of lobes are in-phase.

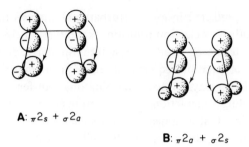

A: $_\pi 2_s + _\sigma 2_a$

B: $_\pi 2_a + _\sigma 2_s$

Conrotatory Opening of Cyclobutene

Figure 4.12

choose, as in **B**, to use the sigma bond suprafacially; this necessitates use of an upper and a lower π lobe, so the π bond is used antarafacially.

The orbital correlation diagram, using the C_2 axis which is preserved throughout the reaction to classify the orbitals, is shown in Figure 4.13. Note that as the ring begins to open σ begins to mix with π^*, which has the same S symmetry. The in-phase combination becomes ψ_2 and the out-of-

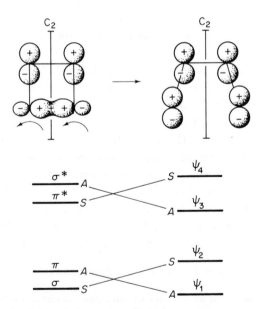

Orbital Correlation Diagram for the Conrotatory Opening of Cyclobutene

Figure 4.13

phase combination ψ_4. The bonding π orbital mixes with σ^* to give ψ_1 from the in-phase combination and ψ_3 from the out-of-phase. Despite the fact that the lowest energy MO in the reactant is not correlated with the lowest in the product, bonding levels only correlate with other bonding levels, so that the reaction is expected to proceed smoothly.

In contrast, the correlation diagram for the disrotatory reaction, using the plane of symmetry preserved throughout the reaction to classify the orbitals, shows that there is an unfavorable correlation, which is just what we expect for a $4n$ system with no, or an even number of, antarafacial components.

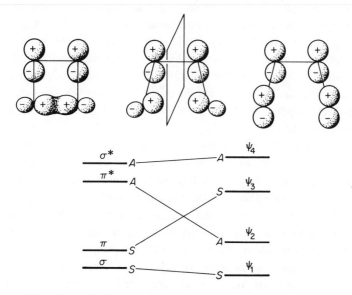

Orbital Correlation Diagram for the Disrotatory Opening of Cyclobutene

Figure 4.14

Problem 4.7. Show that the ring opening of cyclobutene in a disrotatory mode can be analyzed as involving either zero or two antarafacial components.

Problem 4.8. Predict the mode of photochemical ring closure of butadiene.

Problem 4.9. Construct correlation diagrams for, and predict the preferred mode of interconversion of, hexatriene and cyclohexadiene in thermal and photochemical reactions.

Möbius Systems and Generalized Rules
for Pericyclic Reactions

It is interesting to compare the correlation diagrams for the two modes of cyclobutene opening. In addition to the very important fact that just one correlates only bonding MO's in reactant and product, there are several other differences. One is the pattern of energy levels in the transition state. In the conrotatory mode all the energy levels occur in pairs, while in the disrotatory we see the familiar Hückel pattern of the circle mnemonic, which gives rise to the $4n + 2$ rule. Obviously a system which contains an odd number of antarafacial components is not a Hückel system; its pattern of energy levels is different, and we have already observed that it appears to follow a $4n$ rule for stability. Another difference is the correlation of the lowest MO in reactant and product in Hückel systems; whereas, systems that have an odd number of antarafacial components apparently do not have the lowest MO's correlated. What is the reason for this?

In all the Hückel systems we have examined we have always been able to assign phases to the constituent AO's so that it is possible to circumnavigate a pericyclic system without ever encountering a sign inversion. Thus, in a Hückel type transition state for a pericyclic reaction the nodeless lowest MO in the reactant can be transformed into the nodeless lowest MO in the product. In the transition state for a reaction which involves an odd number of antarafacial components, however, it is impossible to traverse the array of orbitals without encountering a sign inversion, as the reader may care to verify for himself by inspecting some of the arrays of this type shown in Figure 4.15. The result is that the nodeless lowest orbital in the reactant acquires a node in a transition state involving such an array; and so not it, but another higher energy orbital, is transformed into the lowest MO of the product.

Arrays of orbitals which cannot be circumnavigated without one, or an odd number of, sign inversions are called Möbius systems. The reason is that a Möbius strip made of overlapping p orbitals has this same property of requiring one sign inversion; and historically, it was the first such system to be investigated theoretically. Heilbronner was able, for this system, to derive the energy levels and the MO's, written as the real and imaginary part of

$$\psi = \sum_r e^{i(c/\lambda)\theta_r} \phi_r \tag{1}$$

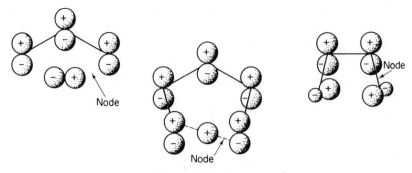

Some Möbius Systems

Figure 4.15

Unlike a Hückel system, the boundary condition in a Möbius strip of $2p$ orbitals is that on traversing 2π radians the wavefunction does not go into itself but into its negative. This requires

$$e^{i(c/\lambda)\theta_r} = -e^{i(c/\lambda)(\theta_r + 2\pi)} \qquad (2)$$

which is satisfied by

$$\frac{2\pi c}{\lambda} = (2m - 1)\pi \qquad (m = 1, 2, \ldots, N) \qquad (3)$$

giving the wavefunctions as the real and imaginary parts of

$$\psi_m = \sum_r e^{i2\pi[m-(1/2)]r/N} \phi_r \qquad (m = 1, 2, \ldots, N) \qquad (4)$$

or

$$\psi_m = \sum_r e^{i2\pi mr/N} \phi_r \qquad (m = \tfrac{1}{2}, \tfrac{3}{2}, \ldots, N - \tfrac{1}{2}) \qquad (5)$$

We can also obtain an analytic expression for the energies by operating on ψ_m with the Hamiltonian, multiplying by ϕ_r, and integrating. As before we obtain, assuming all β equal,

$$E = \alpha + 2\beta \cos\frac{2\pi m}{N} \qquad (6)$$

but unlike (23) of Chapter 3, m can have the values $1/2, 3/2, \ldots, N - 1/2$, or equivalently, $m = \pm\, 1/2, \pm\, 3/2, \ldots \pm 1/2\,(N - 1)$ for N even. Once again a geometric interpretation is possible; for (6) gives the points of intersection between a circle of radius 2β and a regular polygon of N sides which is inscribed in it with *one side down*. The orbital pattern obtained for a 4-membered Möbius system is illustrated in Figure 4.16.

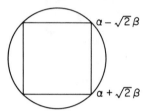

Circle Mnemonic for Möbius Cyclobutadiene

Figure 4.16

Problem 4.10.   Use symmetry to help find the MO's and associated energies for the transition state in the conrotatory cyclobutene ring opening, assuming all β are equal in magnitude.

Problem 4.11.   (a)   Hückel and Möbius systems differ in that in one of the latter type a resonance integral, β, is replaced by $-\beta$ at the point of orbital phase inversion. Repeat the PMO derivation in Chapter 3 of the $4n + 2$ rule for Hückel systems, using Hückel odd AH's and putting an orbital phase inversion at a junction. Show that Möbius systems will follow a $4n$ rule for stability.

(b)   Use Equation (59) of Chapter 3 to derive the rule for finding the coefficients of the starred atoms in the NBMO of a Möbius AH, with particular attention to the coefficient of the starred atom at the point of phase inversion. Repeat the derivation in (a), putting the phase inversion in the odd AH.

We see immediately that a Möbius system will, in contrast to a Hückel system, obey a $4n$ rule in order to make maximum use of its bonding potential and to ensure a smooth transformation of MO's between reactant and product. Thus, Zimmerman was able to formulate a general rule for reactions involving cyclic transition states. If such a transition state has an even number of sign inversions between overlapping orbital lobes (Hückel system), the reaction can proceed readily when

$4n + 2$ electrons are involved. If, however, there are an odd number of sign inversions, the reaction will proceed most easily with $4n$ electrons.

Woodward and Hoffmann have offered another formulation of this general rule for reactions proceeding by monocyclic transition states. For these pericyclic reactions their rule is that "a ground state pericyclic change is symmetry-allowed only when the total number of $(4q + 2)_s$ and $(4q)_a$ components is odd," where q is an integer and the figures in brackets refer to the number of electrons involved in an s (suprafacial) and a (antarafacial) fashion. A component may be a π or group of π bonds, a sigma bond, or a lone AO.

We have suggested that the Woodward-Hoffmann formulation is equivalent to Zimmerman's; let us outline a proof that this is the case. According to the preceding rule, an allowed reaction requires either A: $(2n + 1)(4q + 2)_s$ and $2m(4q)_a$ components or B: $2n(4q + 2)_s$ and $(2m + 1)(4q)_a$. Now an antarafacial component involves one sign inversion.[9] Assuming for the moment that there are no $(4q + 2)_a$ components involved in the reaction, in case A there will be $2m$, an even number, of sign inversions, and a total of

$$8qn + 4q + 4n + 2 + 8mq = 4(2qn + q + n + 2mq) + 2 = 4N + 2$$

electrons. In case B there are $2m + 1$, an odd number, of sign inversions and $8nq + 4n + 8qm + 4q = 4N$ electrons. If there are p additional $(4q)_s$ components, the total number of electrons required for an allowed reaction is $4(N + pq) + 2 = 4N' + 2$ in A, a Hückel system, and $4(N + pq) = 4N'$ in B, a Möbius array. If there are, in addition, an even number, $2p$, of $(4q + 2)_a$ components, the number of sign inversions will remain even in A and odd in B, and $2p(4q + 2) = 8pq + 4p$ electrons will be added to the system, again giving respectively, a total of $4N' + 2$ and $4N'$ electrons. However, if there are an odd number, $2p + 1$, of $(4n + 2)_a$ components, then in case A the number of sign inversions becomes odd, but now there are

$$4N + 2 + (2p + 1)(4q + 2) = 4N + 8pq + 4q + 4p + 4 = 4N'$$

electrons; and in case B the number of inversions becomes even but there

---

[9] In determining whether a given array of orbitals is Hückel or Möbius in nature, it may prove easier to count the number of bonds used antarafacially in the reactant than to draw the orbitals in the transition state and count the sign inversions.

are $4N' + 2$ electrons. Therefore, the Woodward-Hoffmann formulation is equivalent to the statement that a ground state pericyclic reaction is symmetry-allowed only if there are $4n + 2$ electrons in a Hückel or $4n$ electrons in a Möbius transition state.

Of course, these are the prescriptions for "aromaticity" in the Hückel and Möbius systems, so that with greater brevity we may say that pericyclic reactions take place most readily via "aromatic" transition states. As we have seen, in such a transition state molecular orbitals of the reactant are smoothly transformed into those of the product, while an "antiaromatic" transition state, which lacks this smooth correlation, has the electronic structure and energy of a diradical.

Problem 4.12.   By counting the number of electrons in each of the reactions shown in Figure 4.17, you should be able to answer the following questions:

(a)   Why is the product of this cycloaddition formed with *trans* stereochemistry?

(b)   Which geometrical isomer of 2,4-hexadiene will be formed in this reaction? [Hint: Draw the AO's comprising the bonds that are broken in the reactant; and from the mode in which they are required to interact in order to produce an "aromatic" transition state, deduce how the rings will open.]

(c)   The hexamethylbicyclo[3.1.0]hexenyl cation undergoes degenerate thermal [1,4] sigmatropic shifts so rapidly that five of the methyl

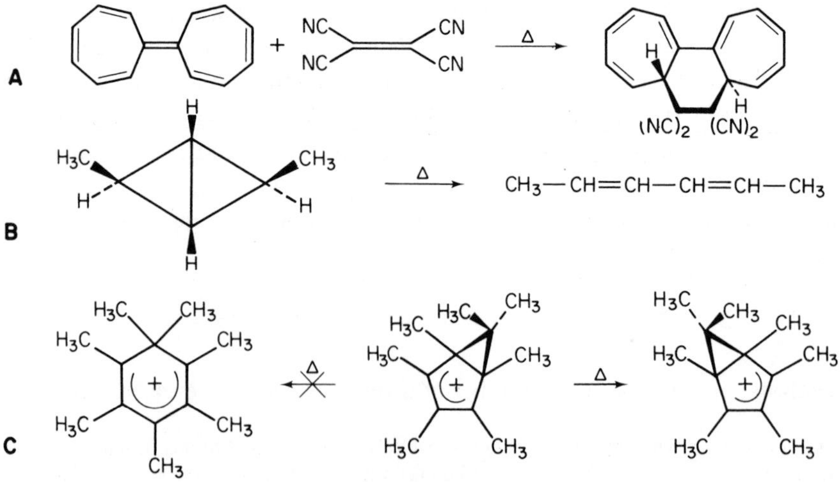

Figure 4.17

groups appear equivalent by nmr. Predict whether the remaining two methyl groups will appear equivalent in the nmr spectrum of this cation. Why does this cation not undergo thermal opening to the cyclohexadienyl cation? Suggest a way of preparing the former from the latter.

Violations

"There are none!" state Woodward and Hoffmann. "Nor can violations be expected of so fundamental a *principle*[10] of maximum bonding." Nevertheless, just as we had to exercise some caution in applying the $4n + 2$ rule to stable molecules, so we must be judicious in our application of the Woodward–Hoffmann rules to transition states.

For instance, the rules do not say anything about diradical reactions except when one may expect them. We have seen that "antiaromatic" transition states for concerted reactions are diradical in nature. If such a transition state is more accessible than one in which a diradical is formed by making or breaking bonds one at a time, then a highly stereospecific reaction may be observed; yet it is still a reaction which does not violate the rules.[11] A much better criterion for testing the rules is comparison of activation energies for two very similar reactions, one of which can proceed by an "aromatic" transition state and one of which cannot. An almost ideal class of such reactions is the electrocyclics, since here the same reactant can go to two nearly identical products (provided steric effects are minimized); and the preferred reaction path should be the one predicted by the rules. This does not mean, however, that even in an ideally designed system, where the only preference for reaction mode is electronic, that the rules are violated if a small amount of the unexpected product turns up. Since the rules are qualitative in nature, they do not predict the quantitative difference in energy between an "aromatic" and a diradical transition state. In fact, in a large system where the highest filled MO is itself close to nonbonding and has small coefficients at the termini, the

[10] The italics and the following interpretation of just what constitutes the Woodward–Hoffmann rules are the author's. The quotation is from R. B. Woodward and R. Hoffmann, *The Conservation of Orbital Symmetry*, Verlag Chemie GmbH, Weinheim, Germany and Academic Press Inc. New York, 1970, p. 173.

[11] The course of reactions that are formally forbidden in the Woodward-Hoffmann sense is an area of current theoretical and experimental research. See, for instance, the work of the Berson group at Yale on sigmatropic reactions, reviewed in *Accts. Chem. Res.* **5,** 406 (1972).

"allowed" and "forbidden" reaction paths might differ little in energy. What would constitute a violation of the rules would be if the disallowed reaction predominated.

Problem 4.13.   Show by constructing a correlation diagram that there exists an allowed pathway for concerted ethylene dimerization, in which one of the ethylenes is used as an antarafacial component. Does the fact that most ethylene dimerizations appear to proceed by diradical mechanisms imply a violation of the Woodward-Hoffmann rules?

Another sure way to get into trouble is to apply the rules for "aromatic" transition states to systems for which they were not designed. For example, in a six electron transition state which consists of a Hückel array of orbitals, prismane opens to benzene. Is the reaction allowed? Careful analysis indicates that it is not, and it is probably to this fact that prismane owes its existence.

Problem 4.14.   Use two planes of symmetry to classify the orbitals and construct a correlation diagram for the benzene–prismane reaction. Show that the reaction fails because the benzene bonding MO with a nodal plane through two atoms fails to correlate with the unique type of bond that is broken in prismane.

The simple analysis of prismane fails to make the correct prediction because the transition state does not consist of the monocyclic array of orbitals for which the rules regarding aromaticity were derived; in fact, the array is tetracyclic. Two of the junctions appear relatively innocuous, since they join atoms belonging to the same set in the benzenoid array, and so might be expected to have little effect on the energy. However, the third is between two *para* positions of the monocyclic ring, and we have seen in the previous chapter that an interaction here is highly destabilizing, since the bond order between these atoms is negative. Thus, if one views the reaction in the direction from benzene to prismane, one is trying to create a sigma bond between two atoms between which the bond order is negative. This suggests that the reaction fails because the benzene MO with nodes through the bonds fails to correlate with the σ orbital. The diagram shows that this, in fact, is not quite the case. The benzene orbital alluded to above actually correlates with an antisymmetric combination of sigma orbitals; it is the failure of the other degenerate benzene bonding MO to correlate with the 1,4 sigma bonding orbital, that results in a crossing. Although first order perturbation theory suggests that the reaction will be difficult; the correlation diagram gives us a more accurate insight into why this is the case.

Effects of Through-Space and Through-Bond Interactions

Another example of the deeper insight into the course of concerted reactions, made possible by construction of a correlation diagram, involves the same 1,4 interaction in a benzenoid-like array. This interaction must exist to some extent in the boat transition state for the Cope rearrangement; and since it is destabilizing, one can readily understand why the chair conformation, in which it is smaller, is preferred. Shown in Figure 4.18 are two resonance structures for the transition state. We have previously alluded to the fact that the lack of clarity as to what phenomenon resonance structures describe limits the usefulness of resonance theory. We now give some definition to what we mean by writing two or more resonance structures for a molecule by specifying that, taken together, they are to represent the bonding in an electronic configuration of a molecule, corresponding to the assignment of electrons to a *single*[12] set of MO's. Therefore, if the two resonance structures in Figure 4.18 are together to portray the partial bonds created through the occupation of certain MO's by the electrons, we must demand that MO's constructed for the localized bonds in the two resonance forms correlate with each other. The correlation diagram is shown in Figure 4.18.

If the orbitals are to correlate smoothly, we see that we must have $p + p'$ above the antisymmetric combination as we have indicated in the diagram. Were the ordering reversed, an orbital crossing would result.

---

[12] The failure of the two resonance structures for cyclobutadiene to satisfy this criterion is why the molecule has no special stability over either one separately. The two resonance structures can be seen to demand that different sets of MO's be occupied, since the $_\pi2_s + _\pi2_s$ reaction, required to convert one into the other, is forbidden. Thus, the MO's of one structure have symmetry different from the MO's of the other, since there is an orbital crossing involved in correlating them. The fact that two equivalent resonance structures for cyclobutadiene can be written actually corresponds to the fact that heavy configuration interaction occurs between them, as we shall see in Chapter 7. By contrast, in benzene the two resonance structures can be interconverted by an allowed $_\pi2_s + _\pi2_s + _\pi2_s$ reaction, and, taken together, they do represent the delocalized bonding which is responsible for special stability of the π system.

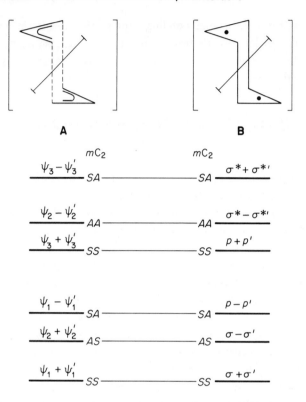

Orbital Correlation Diagram for Two Resonance Structures for the Cope Rearrangement

Figure 4.18

Since the antisymmetric combination $p - p'$ must be filled, there exists an antibonding interaction through space between p and p', which persists in the boat conformation of the transition state. In the boat form the magnitude of β between these two AO's is much larger than in the chair; consequently, the negative bond order between them serves to destabilize the boat with respect to the chair transition state in the Cope rearrangement.[13]

---

[13] Another interesting conclusion that can be drawn from the correlation of the two resonance structures is that radical stabilizing substituents at C-2 and C-5 will lower the energy of the transition state for the Cope rearrangement. If only structure **A**, which pictures the transition state as consisting of interacting allyl radicals, is considered, the prediction that substituents, placed at the carbons where ψ_2 of each allyl radical has a node, will stabilize the transition state is much less easily made. However, when it is realized that the in-phase combination of ψ_2 orbitals is greatly stabilized by the σ

What causes $p - p'$ to lie below $p + p'$, despite the fact that interaction through space should tend to reverse this ordering? Referring to the right hand side of Figure 4.18, we see that the orbital $\sigma + \sigma'$ has the same symmetry as $p + p'$ so they mix. In fact, the in-phase combination is just $\psi_1 + \psi_1'$ and the out-of-phase becomes $\psi_3 + \psi_3'$ as shown in Figure 4.19. Thus, the symmetric combination of p orbitals is destabilized by its interaction with the σ bonding orbitals. In contrast, the antisymmetric combination has symmetry appropriate for mixing with the antibonding σ orbitals which lie above it in energy. As a result it is stabilized, and the mixing transforms it into $\psi_1 - \psi_1'$. Thus, what stabilizes the antisymmetric relative to the symmetric combination of p orbitals is their interaction with the orbitals of the sigma bonds. Hoffmann, who has investigated theoretically such interactions between two nonbonding orbitals, terms this a "through-bond" interaction, to differentiate it from the direct "through-space" interaction of the p orbitals.

In general, we expect that through-space interaction will stabilize

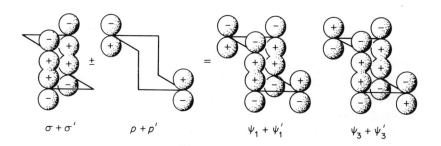

Mixing Between $\sigma + \sigma'$ and $p + p'$

Figure 4.19

---

interaction between the allyl termini, while the out-of-phase combination of ψ_1 orbitals is destabilized, this result is easily understood. The prediction that radical stabilizing substituents at C-2 and C-5 will lower the energy of the transition state is readily made, from **B**, because this structure emphasizes σ rather than π interactions. However, precisely for this reason, it is more difficult to see from **B** than from **A** that substituents at the other four carbons can also stabilize the transition state. Recently DEWAR and WADE [*J. Amer. Chem. Soc.* **95**, 290 (1973)] have found that phenyl substitution at either C-2 or C-3 of 1,5-hexadiene accelerates the rate of Cope rearrangement. It should be noted, however, that Dewar and Wade choose to interpret their experimental results in a way which assigns independent existence to two transition states for the Cope rearrangement with different electronic structures.

the symmetric combination of 1,4 nonbonding AO's. A through-bond interaction will tend to stabilize the antisymmetric combination, since filled bonding sigma MO's will interact with the symmetric combination, resulting in the raising of its energy, while the interaction of the antisymmetric combination with antibonding σ MO's will lower its energy. Which effect predominates depends on geometry. If, in contrast, nonbonding AO's are situated on adjacent carbons, then "nonbonding" is somewhat of a misnomer; for the orbitals interact strongly through space to give a bonding and an antibonding MO. However, when two nonbonding AO's are not nearest neighbors, so that their through-space interaction is small, then the through-bond interaction can dominate, provided, of course, that there is present in the molecule at least one suitably situated sigma bond to interact strongly with the nonbonding orbitals.

Although the interaction of nonbonding orbitals may occasionally have chemical consequences, as in the interaction of the "nonbonding" AO's in the localized σ bond representation of the transition state for the Cope rearrangement, more often the splitting between symmetric and antisymmetric combinations can be observed spectroscopically. Photoelectron spectroscopy, in particular, allows one to measure directly orbital energies.[14] Through analysis of the vibrational fine structure in the photoelectron spectrum of diazabicyclooctane (shown in Figure 4.20), Heilbronner was able to show that the antisymmetric combination of nitrogen nonbonding AO's actually was 2.1 eV (23 kcal/mole-eV) lower in energy than the symmetric combination. In this molecule the through-space interaction is, of course, small; and the lone pairs can interact strongly with the three sigma bonds parallel to them. One might wonder

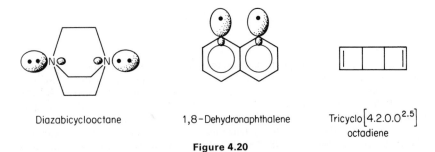

Diazabicyclooctane   1,8–Dehydronaphthalene   Tricyclo$\left[4.2.0.0^{2.5}\right]$
                                                                          octadiene

Figure 4.20

[14] Consequently, of all spectroscopic methods, photoelectron spectroscopy allows the most direct test of the results of MO theory, which accounts for the great current interest in this technique. (See the suggestion for further reading on this subject at the end of this chapter.)

whether one ought also to consider the interactions of the lone pair orbitals with the orbitals of the six other C—C bonds, despite their less favorable orientation. The answer is that these six orbitals occur in pairs, and each pair must be considered to form one combination which is symmetric and one which is antisymmetric with respect to the plane bisecting the molecule. Thus, there will be an equal number of symmetric and antisymmetric bonding orbitals which come from these six bonds; the same is true of the antibonding MO's. The effect of these six sigma bonds on the relative energies of the symmetric and antisymmetric combination of nitrogen orbitals will, therefore, be small compared to the three bonds in which only the symmetric combinations are bonding and the antisymmetric antibonding.

Problem 4.15. Which combination of nitrogen nonbonding orbitals do you expect to lie lowest in pyridazine (1,2-diazabenzene) and in pyrazine (1,4-diazabenzene)?

Problem 4.16. The addition of *o*-benzyne to double bonds is not stereospecific, in conformity with expectations based on the Woodward–Hoffmann rules; but the suprafacial addition of 1,8-dehydronaphthalene to double bonds gives evidence of being concerted. Explain how through-bond interactions in the latter molecule might act to make its cycloaddition reactions "exceptions that prove" the Woodward–Hoffmann rules.

Problem 4.17. As expected, in the photoelectron spectra of *syn*- and *anti*-tricyclo[4.2.0.0^{2,5}]octadiene different ionization potentials are observed for the in-phase and out-of-phase combinations of the bonding π orbitals. However, the energy difference between the ionization potentials of the two combinations is three times larger in the *anti* isomer than in the *syn*. Explain this result and propose a reason, consistent with your explanation, for why the *syn* isomer does not undergo photochemical ring closure to cubane. [Hint: What electronic configuration in two ethylene molecules is required for photochemical cyclobutane ring closure? Might the orbital occupancy of the lowest excited state of the *syn* isomer differ from this?]

FURTHER READING

R. B. WOODWARD and R. HOFFMANN, *The Conservation of Orbital Symmetry*, Academic Press, N. Y., 1970, also appeared in article form in *Angew. Chem., Inter. Edit.*, **8**, 781 (1969). This gives a more complete description than that contained in this chapter and is rich in experimental examples. See also R. E. LEHR and A. P. MARCHAND, *Orbital Symmetry–A Problem Solving Approach*, Academic Press, N. Y., 1972.

H. E. ZIMMERMAN, *Accts. Chem. Res.*, **4**, 272 (1971) is a review of the Möbius-Hückel concept with applications to ground state molecules as well as to transition states.

R. HOFFMANN, ibid., **4**, 1 (1971), presents a discussion of the through-space and through-bond concepts and applications.

H. BOCK and B. G. RAMSAY, *Angew. Chem. Int. Edit.*, **12**, 734 (1973), gives a review of the technique of photoelectron spectroscopy, its applications, and the MO interpretation of the results obtained.

5

Some Failings
of Hückel Theory

The previous chapters have shown that Hückel theory can be very successful, despite the fact that it fails to include electron repulsion explicitly in the Hamiltonian. Nevertheless, we have already encountered several instances where Hückel theory proved to be unsatisfactory. One was its inability to differentiate between the energy of singlet and triplet cyclobutadiene. Another related failure, which we encountered when we discussed diradicals in connection with the Woodward-Hoffmann rules, was the impossibility of doing configuration interaction within the framework of Hückel theory. In this chapter we shall highlight some of its further failings, as motivation for the development in the chapters that follow of a theory which overcomes the shortcomings of the Hückel method, albeit at the expense of some of the simplicity of HMO theory.

Theoretical Critique of the Hückel Method

We expect that Hückel theory will begin to fail to make accurate predictions when electrostatic effects, which it ignores, begin to become important energetically. Thus, we may expect that Hückel theory will not do very well with highly polar molecules where there is substantial charge separation. We need not look far to find a molecule for which Hückel theory makes an incorrect prediction. In Chapter 2 we were able to predict correctly that a water molecule should be bent. We might also predict then that Li_2O should be bent, the only difference being that Li uses $2s$ rather than $1s$ orbitals to form MO's. Experimentally, however, lithium oxide is found to be linear. The reason is that its $2s$ AO's are less electronegative than hydrogen's $1s$ orbitals; therefore negative charge tends to build up more on oxygen, leaving the lithium atoms with significantly larger partial positive charges than the hydrogens in water. All this Hückel theory accounts for, at least in a qualitatively correct manner, through using different α's for the AO's of hydrogen and lithium. What Hückel theory ignores is the fact that because the lithium atoms in the oxide both bear substantial positive charges, they tend to repel each other electrostatically, so that the molecule resists bending.[1]

Even in hydrocarbons, where all the orbitals which constitute the π system have the same α, we have already alluded to the fact that Hückel theory grossly overestimates dipole moments. The reason is that Hückel theory tries to satisfy the $4n + 2$ rule for rings of odd numbers of atoms. Thus, it tends, for instance, to put an additional electron into the π system of a five-membered ring and remove one from that of a three-membered ring. Therefore, it predicts very large dipole moments for methylenecyclopropene, fulvene, azulene, and, as mentioned in Chapter 3, a whopping 15 D, for pentatriafulvalene. The dipole moments that Hückel theory predicts are unrealistically large because in HMO theory it costs nothing, energetically, to separate charge. Although Hückel theory ignores electrostatics, nature, of course, does not. For instance, azulene is predicted by HMO theory to have a dipole moment of 6.9D; the experimental value is 1.0D.[2]

[1] In all fairness to Hückel theory, it does predict a larger energy gap between π_u and $2\sigma_g$ in Li_2O than in H_2O, so that the mixing between these orbitals, which is responsible for H_2O being bent, is expected to be smaller in Li_2O.

[2] The discrepancy between theory and experiment is somewhat lessened if one allows for variation of β with bond length; nevertheless, the theoretical estimate remains about five times too large.

Moreover, even in molecules where the wavefunction is wholly determined by symmetry, Hückel theory errs in that it attempts to write total energies as sums of orbital energies. If, as we have done in Chapter 1, we attempt to account for the increased electron repulsion in an MO wavefunction by adjusting α so that it is not equal to $-I$ but to $-\frac{1}{2}(I + \mathcal{Q})$, then in Hückel theory the total energy of H_2 is

$$E = 2\alpha + 2\beta = 2\beta - (I + \mathcal{Q}) \tag{1}$$

However, in Problem 1.10 we found that when two electrons occupy the same AO, instead of AO's on different atoms at infinity, the electron repulsion energy increases by $I - \mathcal{Q}$. Since in the MO wavefunction for H_2 two electrons occupy the same AO half the time, the energy compared to isolated atoms increases by just half this amount, $\frac{1}{2}(I - \mathcal{Q})$. The total energy of H_2 then is not given by (1) but by

$$E = 2\beta - 2I + \tfrac{1}{2}(I - \mathcal{Q}) = 2\beta - (I + \mathcal{Q}) - \tfrac{1}{2}(I - \mathcal{Q}) \tag{2}$$

Note that Equation (2) differs from Equation (1) by $\frac{1}{2}(I - \mathcal{Q})$, because (1) really counts the increase in electron repulsion, $\frac{1}{2}(I - \mathcal{Q})$, twice, once for each electron. This is, of course, incorrect.

Moreover, even (2) does not really give the correct energy of the hydrogen molecule, for it contains a correction term for the increase in repulsion energy which is based on a model of two atoms at infinity. Clearly, the atoms in the hydrogen molecule are not at infinity. If there were no polar terms in the wavefunction, this might not matter since in two neutral atoms the electron–electron and nuclear–nuclear repulsions should just about cancel the nuclear-electron attractions. However, when both electrons reside on the same atom, as they do with fifty percent probability in the MO wavefunction, not only is there an increase in repulsion between the electrons, for which we have already accounted correctly, but there is also an attraction between the net positive charge at one atom and the net negative charge at the other. This attractive term, which we will call $-\gamma_{AB}$, will certainly not be negligible at the internuclear distance in H_2, but Equation (2) does not yet contain this term for the fifty percent of the wavefunction that places both electrons on the same nucleus. Adding $-\frac{1}{2}\gamma_{AB}$ to (2) and recalling that in Problem 1.10 we found that $I - \mathcal{Q}$ is just γ_{AA}, the repulsion energy of two electrons on the same atom, the total energy of H_2 is given correctly by

$$E = 2\beta - 2I + \tfrac{1}{2}(\gamma_{AA} - \gamma_{AB}) \tag{3}$$

Now although we might consider redefining α as $-I + \frac{1}{4}\gamma_{AA}$ and including $-\frac{1}{4}\gamma_{AB}$ in β, so we can be consistent with (3) and still write total energies as sums of orbital energies,[3] some reflection shows us how complicated this can become, even for the H_2 system. For instance, in the hydrogen molecule ion, H_2^+, there is no electron repulsion, and so a different α and β from the ones we use for neutral H_2 would be needed. And what would we use for α and β in H_2^-?

Clearly, we can expect Hückel theory to work best in calculating relative energies of neutral hydrocarbons because here we never really need to specify what α is. In calculations on this type of molecule, we can write total energies as sums of orbital energies and *hope* that electron repulsion will be taken account of in some average way in the adjustable parameter β. Since we know that we cannot expect good wavefunctions from HMO theory for molecules containing rings consisting of odd numbers of atoms, we conclude that we can expect Hückel theory to work best for the neutral AH's—the class of molecules for which we were able to obtain reasonably good results in Chapter 3. When we have developed a theory which explicitly includes electron repulsion in the Hamiltonian, we will see in some more detail why HMO theory is successful in its treatment of this class of compounds.

Just because Hückel theory works well in the calculation of relative energies of neutral AH's, does not mean that it succeeds in predicting other properties of this class of molecules. In the next two sections of this chapter we will consider the electronic absorption and electron paramagnetic resonance spectra of some AH's. We shall find that although Hückel theory is successful in describing some aspects of these spectra, it fails to give a full account of all the features that are observed, precisely because of the omission from the Hamiltonian of the electron repulsion operator.

Electronic Spectra of Alternant Hydrocarbons[4]

The absorption of electromagnetic radiation is a time-dependent phenomenon; so we must use the time-dependent Schrödinger equation

---

[3] We will see in Chapter 8 that doing this creates some formal resemblance between the expressions for the total π energy of a neutral AH in HMO and in a more complete theory that explicitly includes electron repulsion in the Hamiltonian.

[4] A mastery of the mathematical material in this section is not essential to the comprehension of later chapters, nor, indeed, to the understanding the qualitative concepts presented here.

$$[\mathcal{H}(r) + \mathcal{H}'(r, t)]\Psi(r, t) = -\frac{\hbar}{i}\frac{\delta\Psi(r, t)}{\delta t} \tag{4}$$

As usual, we proceed from the known to the unknown and write

$$\Psi(r, t) = \sum_m c_m(t)\Psi_m(r, t) \tag{5}$$

where $\Psi(r, t)$ is the time-dependent wavefunction we seek, the $\Psi_m(r, t)$ are the solutions of (4) when the Hamiltonian is time-independent, and the $c_m(t)$ are the time-dependent coefficients we must obtain. The method of solving the problem should be familiar by now; we multiply (5) sequentially by $\Psi_n^*(r, t)$ and integrate over space.[5] This gives a series of differential equations of the form

$$\sum_m c_m(t) \int \Psi_n^*(r, t)\mathcal{H}'(r, t)\Psi_m(r, t)\, d\tau = -\frac{\hbar}{i}\frac{\delta c_n(t)}{\delta t} \tag{6}$$

These are usually simplified by assuming that at zero time we know that the system is in a state $\Psi_m(r)$, so that for times near zero all the terms in the summation are negligible, except for the mth, where $c_m(t) = 1$. The $\Psi_m(r, t)$ are easily obtained from the time-independent wavefunctions, $\Psi_m(r)$, which we already know how to find. When the Hamiltonian is time-independent, $\Psi_m(r)$ satisfies the time-independent Schrödinger equation

$$\mathcal{H}(r)\Psi_m(r) = E_m\Psi_m(r) \tag{7}$$

We can obtain (7) from (4) when $\mathcal{H}'(r, t) = 0$ with

$$\Psi_m(r, t) = \Psi_m(r)e^{-iE_mt/\hbar} \tag{8}$$

as the reader may verify by substitution.

Let us assume that at some time t we have determined the mixing coefficients to be $c_1(t)$ and $c_2(t)$ for two states $\Psi_1(r)$ and $\Psi_2(r)$ between which an electronic transition is occurring. Then at time t the wave-

---

[5] Ψ^* is the complex conjugate of Ψ, which we use in the event that Ψ has an imaginary part.

function for the system undergoing a transition is

$$\Psi(r, t) = c_1(t)\Psi_1(r)e^{-iE_1t/\hbar} + c_2(t)\Psi_2(r)e^{-iE_2t/\hbar} \tag{9}$$

Let us now compute the probability $\Psi^*\Psi$ of finding the system in each stationary state. On multiplying (9) by its complex conjugate we obtain

$$\begin{aligned}\Psi^*(r, t)\Psi(r, t) = c_1^2(t)\Psi_1^2(r) + c_2^2(t)\Psi_2^2(r) \\ + 2c_1c_2\Psi_1(r)\Psi_2(r)\cos\left[\frac{2\pi}{h}(E_2 - E_1)t\right]\end{aligned} \tag{10}$$

which gives the relative probabilities of finding the system in state 1, in state 2, and having a transition density $\Psi_1(r)\Psi_2(r)$ which oscillates with a frequency of $(E_2 - E_1)/h$.

Problem 5.1.   Consider Ψ_1 as the ground state of ethylene, $\pi(1)\pi(2)$, and Ψ_2 as the lowest excited state, $\pi(1)\pi^*(2)$, where π is the bonding MO, π^* the antibonding MO, and the numbers in parentheses identify the two electrons.[6] A dipole moment is calculated from

$$\boldsymbol{\mu} = e \int \Psi^* \mathbf{r}\, \Psi\, d\tau \tag{11}$$

Use (10) to calculate the dipole moment for the wavefunction of equation (9). Show that a dipole moment is induced in ethylene, as it undergoes an electronic transition, despite the fact that neither the ground nor excited state has a dipole moment. Show that the induced dipole oscillates with frequency $(E_2 - E_1)/h$ and has an amplitude proportional to

$$e \int \Psi_1 \mathbf{r}\Psi_2\, d\tau = e \int \pi^* x \pi\, d\tau$$

[6] Since electrons are identical, we cannot really label them in this way, which distinguishes between them. In a subsequent chapter we will investigate how to write correct many-electron wavefunctions. For our present purpose the above wavefunctions are adequate.

which is called the "transition dipole moment." [Hint: The vector

$$\mathbf{r} = \mathbf{r}(1) + \mathbf{r}(2)$$
$$= \mathbf{x}(1) + \mathbf{y}(1) + \mathbf{z}(1) + \mathbf{x}(2) + \mathbf{y}(2) + \mathbf{z}(2)$$

operates on only one electron at a time.]

Now we know from physics that a dipole produces electromagnetic radiation of the frequency at which it oscillates, so that associated with an electronic transition between Ψ_1 and Ψ_2 will be radiation of energy $E = h\nu = E_2 - E_1$, provided, of course, that a transition dipole exists. By the principle of time reversibility[7] not only is an oscillating dipole required for the production of electromagnetic radiation; but in order for absorption to occur, an electromagnetic field also must be capable of creating an oscillating dipole in a molecule. Therefore, in order for absorption of radiation to take place, producing an electronic transition between the stationary states Ψ_1 and Ψ_2, the frequency of the radiation must be $(E_2 - E_1)/h$, and the transition dipole moment $\mathbf{\mu}_{12} = e \int \Psi_1 \mathbf{r} \Psi_2 \, d\tau$ must not be zero. The strength of an electronic transition, as measured by the extinction coefficient, is, in fact, proportional to the square of the transition dipole moment. When a transition dipole moment is calculated to be zero, the transition is said to be "forbidden"; however, this does not mean that such a transition may not still be observed to occur with appreciable intensity. The occurrence of a "forbidden" transition usually comes about through a distortion of the molecular symmetry by an anti-symmetric vibration, which mixes excited states and results in a nonzero transition dipole moment. It is also possible for a molecule to absorb electromagnetic radiation by other than the electric dipole mechanism; however, magnetic dipole and electric quadrapole transitions are several orders of magnitude weaker.

Under what circumstances will a transition dipole moment not be zero? When we have learned to write wavefunctions correctly, we shall be able to show that only transitions between states of the same multiplicity have any appreciable dipole associated with them. Thus, singlet–singlet

---

[7] Most simply stated, this principle requires that a film of a microscopic event correspond to physical reality whether it be run forward or backward. Thus, light emission must be just the time reverse of the absorption process. Note the time reversal symmetry manifested in (10), since the cosine function assumes the same value at plus or minus t.

and triplet–triplet electronic transitions can be radiation-induced; singlet–triplet transitions are generally far weaker and occur most efficiently by radiationless mechanisms.[8] Another "selection rule" is that only one-electron excitations can occur with any high degree of probability. Therefore, two states between which an allowed transition can occur must differ only in the assignment of one electron to a different orbital, say ψ_i for the ground state and ψ_j for the excited state. Finally, since $\mathbf{r}$ operates on only one electron at a time, the equation for the transition dipole moment between two states can be broken up into n terms, one for each electron, all but one of which will be of the form

$$\int \psi_1^2(1)\, d\tau \int \psi_1^2(2)\, d\tau \, \ldots \int \psi_i(m)\psi_j(m)\, d\tau \, \ldots \, e \int \psi_k(n)\mathbf{r}(n)\psi_k(n)\, d\tau$$

These terms are all zero, since $\int \psi_i(m)\psi_j(m)\, d\tau = 0$. The equation for the transition dipole moment created by the excitation of an electron from ψ_i to ψ_j thus reduces to the one remaining term[9]

$$\begin{aligned}
\mathbf{\mu}_{ij} &= \int \psi_1^2(1)\, d\tau \, \cdots \int \psi_k^2(n)\, d\tau \, \cdots \, e \int \psi_i(m)\mathbf{r}(m)\psi_j(m)\, d\tau \\
&= e \int \psi_i \mathbf{r} \psi_j\, d\tau
\end{aligned} \tag{12}$$

which may or may not be zero depending on the symmetry of ψ_i and ψ_j. If the integral in (12) is not to be zero, one of the terms in the integrand must be a symmetric function of all three coordinates.[10] Since $\mathbf{r}$ is an antisymmetric function (each of its components, $\mathbf{x}$, $\mathbf{y}$, and $\mathbf{z}$, changes sign on passing through the origin of the molecular coordinate system), this

---

[8] This is true for molecules containing only light atoms. When heavy atoms are present, spin-orbit coupling can cause mixing of wavefunctions of different multiplicity.

[9] Note that if ground and excited states differ in the orbital assignments of two electrons, this term will also involve an integral of the form $\int \psi_k(n)\psi_l(n)$ which is zero by orthogonality. It is for this reason that two-electron excitations are "forbidden."

[10] The integral of an antisymmetric function is always zero, while that of a symmetric function need not vanish. Compare the integrals $\int_{-a}^{a} x\, dx$ and $\int_{-a}^{a} x^2\, dx$. We shall see in Chapter 6 that, more generally, integrals, in order to be nonzero, must be symmetric to all symmetry elements present in a molecule. For the transition dipole moment $\mathbf{\mu}_{ij} = e \int \psi_i \mathbf{r} \psi_j\, d\tau$, this requires that the product $\psi_i \psi_j$ have the same symmetry behavior as one or more of the components of $\mathbf{r}$.

means that the product $\psi_i\psi_j$ must also have a component that is antisymmetric in one coordinate, if the integral in (12) is to have a nonzero value. Physically, this just requires that the transition charge density, $e\psi_i\psi_j$, have a net dipole moment.

Problem 5.2. Which transitions are allowed in *s-trans-* and *s-cis-*butadiene and which are forbidden? [Hint: Sketch the charge distribution given by $e\psi_i\psi_j$ for each $\psi_i \rightarrow \psi_j$ transition, and see which give rise to nonzero transition dipole moments.]

Problem 5.3. Calculate the transition dipole moment $\mu = e \int \pi^* \mathbf{r} \pi \, d\tau$ in ethylene in units of e, the charge of an electron. Assume that atomic electron distributions like ϕ_1^2 and ϕ_2^2 may be replaced by point charges at the position of each carbon nucleus for calculating the dipole moment.

Using Hückel theory we cannot expect to make very accurate predictions about the excited states of molecules. The reason again is that because the electron repulsion operator is not explicitly included in the Hamiltonian, Hückel theory is incapable of differentiating between singlet and triplet states. The energy difference between these two types of excited states comes from the smaller electrostatic repulsion in the latter, and this difference is by no means a negligible effect. In ethylene the difference in energy between these two excited π states is 3 eV or about 70 kcal/mole. It is possible, however, to account for some general features of spectra using Hückel theory. For instance, we have already been able to predict correctly that butadiene should absorb at longer wavelengths than ethylene. One can, provided one allows for bond alternation, even predict the quantitative dependence on chain length of the wavelength and intensity of the first allowed absorption in a series linear polyenes.

There are, however, a number of spectral features that unmodified Hückel theory is incapable of even qualitatively reproducing. For example, it is well known that *cisoid* butadienes absorb energy at considerably longer wavelengths and with smaller extinction coefficients than their *transoid* isomers. The β_{14} resonance integral is too small to account for the roughly 15 kcal/mole difference between them in the energy of the lowest transition, and we may suspect that the difference may arise from an electrostatic interaction that Hückel theory ignores. Indeed, it does; and although this difference arises naturally in a theory which explicitly includes electron repulsion in the Hamiltonian, we have to make some further assumptions in order to account for it in Hückel theory.

Suppose that butadiene can be treated as two ethylenes. Since light absorption would occur at the same frequency in both units, we must imagine that an excitation is localized in neither but is continually being transferred from one ethylene to the other. Both, then, are repeatedly undergoing transitions so that transition dipoles are simultaneously present along both ethylenic units. These dipoles can interact electrostatically, and the effect of their interaction depends on their relative orientation ("head to head" or "head to tail"). The orientation of the transition dipoles along the two ethylenes for the $\psi_2 \longrightarrow \psi_3$ transition can be deduced from the transition charge density, $e\psi_2\psi_3$, for this excitation in butadiene. The transition charges and dipoles are shown below in the *cisoid* and *transoid* molecules, and the drawing reveals the origin of the difference in wavelength and intensity of absorption in the two conformers.

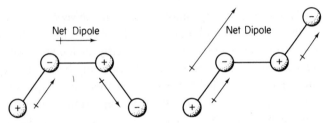

Transition Charges and Dipoles for a $\psi_2 \longrightarrow \psi_3$ Excitation in *Cisoid* and *Transoid* Butadiene

Figure 5.1

Cisoid butadiene differs from the *transoid* isomer in that the dipoles along the 1-2 and 3-4 ethylenic bonds are more favorably disposed for a stabilizing "head to tail" interaction in the former. Thus, *cisoid* butadiene should absorb at longer wavelengths (lower energy). Also, in the *cisoid* isomer the transition dipoles along these bonds tend to partially cancel each other out; whereas in the *transoid* form they add. Therefore, the latter, having a larger net transition dipole moment, is expected to absorb more strongly than the former.

Problem 5.4.   Consider the two double bonds in norbornadiene shown in Figure 5.2. A wavefunction for the excited state must be written as $\Psi = \pi_1^2\pi_2\pi_2^* \pm \pi_1\pi_1^*\pi_2^2$ (the subscripts refer to the two ethylenic units), since the ethylenic units are identical; and from symmetry, the excitation cannot be said to reside in either separately. Sketch the interacting transition dipoles. Which combination, plus or minus, will have the lower energy? the higher extinction coefficient?

Norbornadiene

Figure 5.2

The ad hoc treatment of excited butadiene as two isolated ethylenes that exchange electronic excitation does account for the differences in the spectra of *cisoid* and *transoid* conformations of the system. Nevertheless, this treatment is really not very satisfactory, especially since we know electron delocalization between the two ethylenic units to be very important in the excited state of butadiene. However, in a molecule that, unlike butadiene, is actually predicted by HMO theory to have two or more transitions of the same energy, exchange of the excitation between them with consequent interaction of their transition dipoles is not a bad physical model. For example, were there no through-space or through-bond inter- action between the ethylene MO's in norbornadiene (Problem 5.4), the "exciton" exchange model would give an excellent account of the spectrum of this molecule. The addition of "exciton" exchange to Hückel theory is particularly important in obtaining a reasonable account of the spectra of molecules in which the consequences of the pairing theorem result in transitions that are degenerate in energy in unmodified HMO theory.

Let us examine the case of naphthalene, an even alternant, which from the pairing theorem can be expected to have the pattern of energy levels shown in Figure 5.3. There are three different transitions of relatively low energy, of which $\psi_5 \longrightarrow \psi_6$ is the lowest. The excitations $\psi_4 \longrightarrow \psi_6$ and $\psi_5 \longrightarrow \psi_7$ are of equal energy according to the pairing theorem. Un- modified Hückel theory would lead us, therefore, to expect to observe two absorptions in the naphthalene spectrum; in fact, three are found— a very weak one at longest wavelength followed by a more intense band, with a very intense absorption appearing at even shorter wavelengths. Inspection of the transition charges, $e\psi_i\psi_j$, shown in Figure 5.3 for these three excitations provides the clue to the interpretation of the observed spectrum.[11]

---

[11] Although in Figure 5.3 nodal planes and the transition charges between them have been indicated, for the purpose of determining the orientation of the transition dipoles it is sufficient just to multiply together the coefficients of the two MO's at the atoms where they both have nonzero amplitudes.

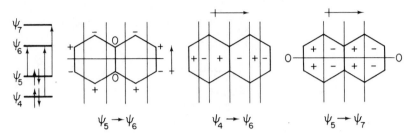

$\psi_5 \rightarrow \psi_6$ $\psi_4 \rightarrow \psi_6$ $\psi_5 \rightarrow \psi_7$

Transition Charges and Dipoles for Low–Lying Excitations in Naphthalene

Figure 5.3

From Figure 5.3 we see that while the lowest energy transition is polarized along the short molecular axis, the transition dipoles for the two excitations with of the same HMO energy both lie along the long molecular axis. Thus, as the "exciton" is transferred back and forth between the degenerate transitions, the transition dipoles can interact. It can be shown from the pairing theorem that the two transition dipoles are of equal magnitude, so we can predict that the out-of-phase combination is a "forbidden" transition and hence will have a small extinction coefficient. Since this combination involves a stabilizing "head to tail" interaction between the dipoles, we can thus identify it as giving rise to the weak, long wavelength band in the spectrum. The transition in which the dipoles are in-phase should be especially strong and of high energy, since it involves a "head to head" interaction of the dipoles; therefore, it is probably the absorption at shortest wavelength in the spectrum.

Problem 5.5. How many low-energy transitions would you expect to see in benzene if all electrostatic effects are neglected? Three bands are in fact observed—two relatively weak ones at long wavelength and one intense band at shorter wavelength. Sketch the benzene MO's and the transition dipoles for the four lowest energy excitations, and qualitatively interpret the benzene spectrum. [Hint: Only three bands are observed because the two allowed transitions turn out to have the same energy and transition dipoles which are orthogonal, so that they cannot interact. The degeneracy of these transitions, even when electrostatic effects are included, is a consequence of molecular symmetry, as we shall see in Chapter 7.]

Reduced to its essentials, the interacting dipole model predicts strong electrostatic interaction between two excited singlet electronic configurations with the same Hückel energy. This is indeed the case; for as we shall see in Chapter 7, excited singlets have wavefunctions in which ionic terms are heavily weighted.[12] Mixing between excited singlet configurations is important for removing this ionic character; and of course, mixing will be especially strong when two electronic configurations have the same energy. In the absence of a Hamiltonian that explicitly includes the electron repulsion operator, the interacting transition dipole model provides a rationale for why such mixing occurs. Nevertheless, the physical picture on which this model is based is not a wholly accurate one, and it can be somewhat misleading if it is naively applied. For instance, two excited configurations need not have the same Hückel energy for mixing between them to occur. Also, as we shall see in Chapter 7 (Problem 7.10), unless two excitations occur in different parts of a molecule (i.e., in two chromophores between which overlap is negligible), there is another term present in the expression for the interaction of the two excited configurations, in addition to the one for electrostatic interaction of their transition dipoles.

Electron Paramagnetic Resonance (EPR) Spectra of Alternant Hydrocarbons

The interaction between the magnetic dipole created by the spin of an unpaired electron and an external magnetic field results in a difference between the energy of an electron with spin up (α) and spin down (β). Transitions between the two levels can be induced by absorption of energy from an oscillating electromagnetic field whose frequency satisfies

$$h\nu = g\beta H \tag{13}$$

where $g = 2.0023$, β is the Bohr magneton $= e\hbar/2mc = 0.97 \times 10^{-20}$ erg/gauss, and H is the magnetic field strength.

In addition to the external field an unpaired electron is also affected by the nuclear spins in the molecule, in particular those of the hydrogen nuclei in hydrocarbons. This type of spin interaction is called hyperfine coupling, and it gives rise to epr spectra consisting of many lines. In solu-

---

[12] The wavefunction for the lowest singlet to which a transition is allowed in butadiene results in C-1 and C-4 being oppositely charged in this excited state. Since the *cisoid* conformation causes less charge separation than the *transoid*, the former is of lower energy.

tion the nuclear spins can affect an unpaired electron only by Fermi contact coupling, which requires that the wavefunction of the unpaired electron have a finite value at the nuclei. Now if a molecule contains a methyl group, the orbitals of which overlap with a p orbital that forms part of a π MO containing the unpaired electron, this overlap provides a mechanism by which unpaired spin density can appear at the hydrogen nuclei of the methyl. However, there would appear to be no mechanism by which a proton attached directly to an sp^2 carbon atom, and hence lying in the nodal plane of a p orbital, could produce hyperfine splitting. Nevertheless, such hyperfine splitting is observed; and it provides a great deal of information regarding the distribution of an unpaired electron in a molecule, which can be related to the wavefunction of the orbital which the electron occupies.

Problem 5.6.   Consider a methyl group attached to an sp^2 carbon in a conformation such that the C—H bond to one of the methyl hydrogens lies in the nodal plane of the p orbital. Show that one of the symmetry-correct combinations of the other two C—H orbitals has the correct symmetry for overlap with the p orbital. (Methyl groups usually rotate so rapidly that all three protons appear to interact equally with the unpaired electron on the time scale of an epr measurement.)

The nonzero hyperfine splitting caused by protons fixed in the nodal plane of a π system, although inexplicable in Hückel theory, is quite understandable when one takes into account electron repulsion. Consider an electron with α spin in a p orbital on a carbon atom to which a proton is attached. If we partition the electrons in the C—H σ bond between the sp^2 hybrid on carbon and the hydrogen $1s$ orbital, there are two different situations possible with respect to the spins of these electrons, as shown in Figure 5.4. When there is no net spin density in the π orbital, **A** and **B**

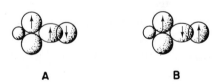

A   **B**

Possible Spin Distributions in the C—H Bond of a Carbon π Radical

Figure 5.4

are of the same energy; but with net α spin in the π orbital, **A** is more stable. The reason is that since the total electronic wavefunction, Ψ, correlates the motions of electrons with the same spin so that they tend to avoid each other, the electrostatic repulsion between the π and σ electrons is slightly smaller in **A** than in **B**. Thus, structure **A** will contribute to the C—H bond slightly more than **B**, and a net spin density will be produced at the hydrogen nucleus. We note that the spin density so produced will be negative in that it is opposite to that in the p orbital; and the hyperfine coupling constant can, therefore, be predicted to be negative in sign, as experimentally it is found to be.

The analysis of the hyperfine splitting in an epr spectrum is very similar to the first-order analysis of the spin–spin splitting in nmr. If an unpaired electron interacts with n equivalent nuclei, $n + 1$ lines will be observed whose relative heights are given by the coefficients in the binomial expansion of $(1 + x)^n$. For instance, in the methyl radical an unpaired electron interacts with three protons, and four lines are observed in the ratio $1 : 3 : 3 : 1$. The separation between any two is 23 gauss, which is the hyperfine coupling constant in the methyl radical.[13] Spectral analysis is not always so easy, however. For instance, in the triphenylmethyl radical there are 3 *para*, 6 *meta*, and 6 *ortho* protons. Since the protons of a given type are equivalent, the spectrum consists of $4 \times 7 \times 7 = 196$ lines.

If the splitting caused by a proton in the methyl radical, where the p orbital contains one unpaired electron, is 23 gauss, we might wonder if the splitting caused by a hydrogen nucleus attached to a carbon where an average of $1/n$ unpaired electrons is localized would be $23/n$ gauss. In fact, in the benzene radical anion the observed hyperfine coupling constant is 3.75 gauss, very close to the $23/6$ gauss predicted for one electron divided equally between six atoms. McConnell has proposed the relation

$$a_r = \rho_r Q \tag{14}$$

which allows the estimation of the electron density ρ_r at atom r from the observed hydrogen hyperfine coupling constant a_r. From one system to another Q is not really constant; for instance, in the cyclooctatetraene radical anion it apparently is closer to 26 than to 23 gauss. However, Equation (14) is qualitatively good enough to allow its use for the inter-

---

[13] Actually, this is the magnitude of the coupling constant; its sign is negative. Since only the magnitudes of hyperfine coupling constants can be obtained directly from epr spectra, only these are given in the text.

pretation of the spectra of a wide variety of radical ions in even alternant hydrocarbons.

Problem 5.7. Sketch the two possible spectra you would expect to observe if the odd electron in the benzene radical anion occupied only one of the two degenerate antibonding MO's. Use (14) to indicate the quantitative splittings between lines. What are the results of working the problem for the radical cation? Explain why the spectrum of the benzene radical anion consists of seven lines each separated by 3.75 gauss.

One way to force the unpaired electron to occupy selectively one of the degenerate antibonding benzene MO's in the radical anion is to substitute the ring so as to lift the degeneracy. Now methyl groups are known to be electron-donating substituents, so that the extra electron in the benzene ring will tend to avoid a carbon to which a methyl is attached. Thus, the electron will prefer to occupy an MO which has a node through that atom. The hyperfine coupling constants in gauss for toluene and *p*-xylene negative ions are shown in Figure 5.5. The smaller spin density on the methyl protons in the *p*-xylene ion suggests that two methyl groups are probably better than one at lifting the degeneracy of the orbitals.

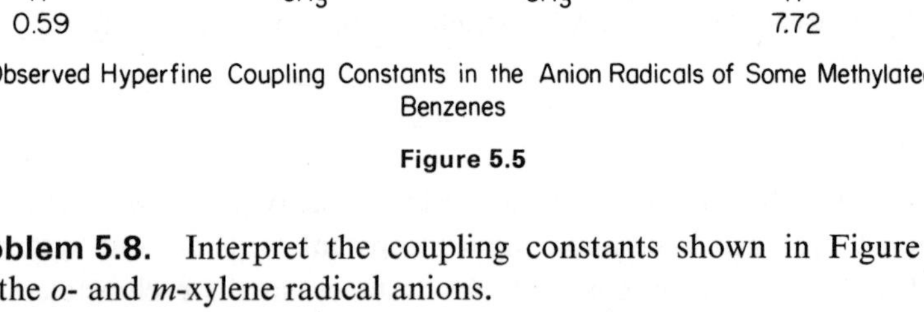

Observed Hyperfine Coupling Constants in the Anion Radicals of Some Methylated Benzenes

Figure 5.5

Problem 5.8. Interpret the coupling constants shown in Figure 5.5 for the *o*- and *m*-xylene radical anions.

The above results indicate that the form of the MO's in benzene is correct. Epr experiments also provide evidence for the correctness of the pairing theorem. Shown in Figure 5.6 are the observed hyperfine coupling constants for the anthracene anion and cation radicals. The pairing theo-

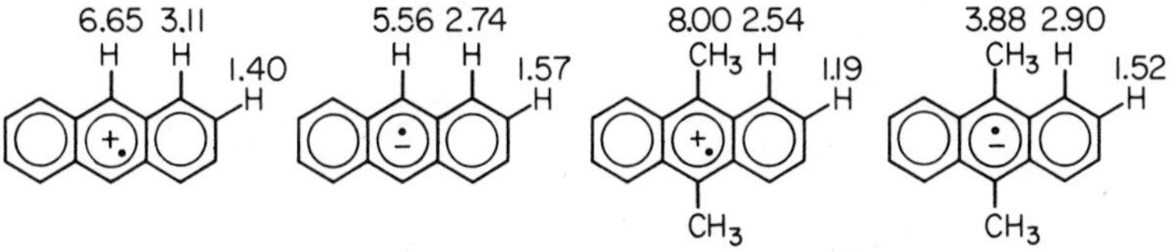

Observed Hyperfine Coupling Constants in the Radical Ions of Anthracene and 9,10–Dimethylanthracene

Figure 5.6

rem does not require that the observed hyperfine coupling constants should be identical in both, for this depends on the constancy of Q in (14). All the pairing theorem requires is that the spin densities be the same in radical anion and cation, since the orbitals occupied by the odd electron are paired. The ratio of spin densities in the cation is 0.9 : 2.0 : 4.2, and in the anion 1.1 : 2.0 : 4.0. Not only are the observed ratios in the two ions very similar, but they are also both very close to the 1 : 2 : 4 ratio predicted by a Hückel calculation.

Problem 5.9. Are the different methyl proton coupling constants in the anion and cation radical of the dimethylanthracene shown in Figure 5.6, a violation of the pairing theorem? Explain qualitatively why a difference is observed.

Hückel theory does well in predicting the epr spectra of even alternant radicals, but it fails badly for odd alternants as shown in Figure 5.7. Not only do the observed hyperfine coupling constants add up to far more than 23 gauss, but spin density appears at positions where we expect none. Once again this failure of Hückel theory can be traced to its failure to deal explicitly with electron repulsion. The reason for the appearance of spin density where we expect none is very similar to the reason for the non-zero value of the hyperfine coupling constant for a hydrogen in the nodal plane of a π system. Again the unexpected coupling is produced by a negative spin density. For instance, in the case of allyl the electron in ψ_1 that has the same spin as the unpaired electron in the nonbonding MO, will, on the average, spend more time on the terminal atoms than does the electron in ψ_1 of the opposite spin. The reason, once more, is that electrons of the same spin have well-correlated motions; in fact, as we shall see in Chapter 7, two π electrons of the same spin cannot appear simultaneously at the same atom. Since the unpaired electron in ψ_2 is confined to the

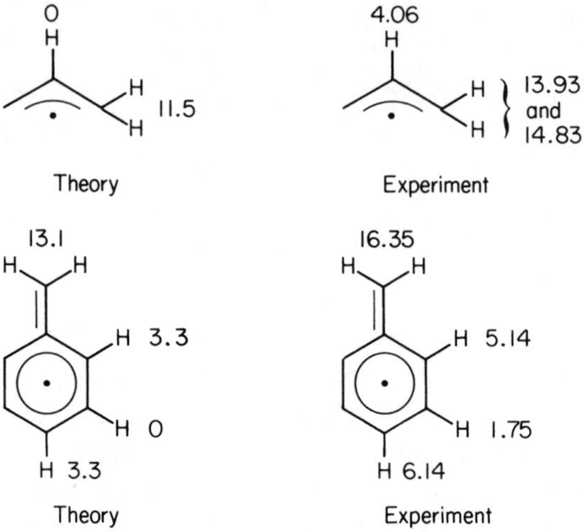

Comparison of Theoretical and Experimental Hyperfine Coupling
Constants in Allyl and Benzyl Radicals

Figure 5.7

termini, electron repulsion is minimized when the electron of the same spin in ψ_1 spends more time on the terminal atoms; for when the electron in ψ_2 is at one terminal atom, the electron of the same spin in ψ_1 tends to be at the other end of the molecule. In order to avoid a net transfer of charge from the central to the terminal atoms, the electron in ψ_1 of opposite spin from that in ψ_2 must spend more time on the central atom. This also tends to keep these two electrons from appearing simultaneously at a terminal atom—an event which is not prohibited for electrons of opposite spin and which results in substantial electrostatic repulsion between them. The effect of the two electrons in ψ_1 having the different wavefunctions is that the positive spin density is increased on the terminal atoms and a negative spin density is produced on the central carbon. That this interpretation is reasonable can be seen from the fact that if the spin density of the central carbon in allyl is considered negative, we obtain a value of

$$Q = 2 \times \tfrac{1}{2}(13.93 + 14.83) - 4.06 = 24.7 \text{ gauss}$$

which is much closer to the value of Q found in the methyl radical and benzene radical anion.

We could, of course, try to correct the Hückel theory description of odd alternant radicals *a posteriori*, as we did with the HMO treatment of u.v. spectra. In this case, however, we will defer any further discussion of negative spin densities to Chapter 9. When we have included electron repulsion explicitly in the Hamiltonian, we will see how negative spin densities in radicals arise naturally.

FURTHER READING

J.N. MURRELL, *The Theory of the Electronic Spectra of Organic Molecules*, John Wiley & Sons, New York, 1963. Chapters 5–7 treat in detail the aspects of the electronic spectra of organic molecules which we have considered.

ALAN CARRINGTON and ANDREW D. MCLACHLAN, *Introduction to Magnetic Resonance*, Harper & Row, New York, 1967. Chapter 6 is concerned with the epr of organic radicals in solution.

6

A More Rigorous
Treatment of Operators
and Their Eigenfunctions—
Group Theory

In the previous chapter we have explored some of the shortcomings of Hückel theory, caused by its failure to include electron repulsion explicitly in the Hamiltonian. In this chapter we will lay the theoretical groundwork necessary for the construction of a theory which overcomes this deficiency and yet retains much of the simplicity of the Hückel method. At the same time we will take this opportunity to make more rigorous, and hence more generally useful, some of the intuitive ideas, especially those regarding symmetry, that we have hitherto applied chiefly on the grounds that they seemed physically reasonable.

Eigenvalue Problems

The Schrödinger equation belongs to a general type known in mathematics as eigenvalue equations. The general eigenvalue problem is, given an operator G, to find a set of eigenfunctions, Γ_m, such that the equation

$$G\Gamma_m = g_m\Gamma_m \tag{1}$$

is satisfied, where g_m is a constant associated with each Γ_m. The g_m are often referred to as eigenvalues. It can be shown that if G is an Hermitian operator, the eigenfunctions found from solution of (1) are orthogonal. Mathematically,

$$\int \Gamma_n^* \Gamma_m \, d\tau = 0 \tag{2}$$

where Γ_m and Γ_n are eigenfunctions of G with associated eigenvalues g_m and g_n, and Γ_n^* is the complex conjugate of Γ_n. Equation (2) can be derived starting from the definitions

$$G\Gamma_m = g_m\Gamma_m \quad \text{and} \quad G\Gamma_n = g_n\Gamma_n \tag{3}$$

Multiplying the first equation in (3) by Γ_n^* and integrating gives

$$\int \Gamma_n^* G\Gamma_m \, d\tau = g_m \int \Gamma_n^* \Gamma_m \, d\tau \tag{4}$$

Taking the complex conjugate of both sides of the second, multiplying by Γ_m and integrating yields

$$\int \Gamma_m G^* \Gamma_n^* \, d\tau = g_n^* \int \Gamma_m \Gamma_n^* \, d\tau \tag{5}$$

Subtracting (5) from (4) we obtain

$$\int \Gamma_n^* G\Gamma_m \, d\tau - \int \Gamma_m G^* \Gamma_n^* \, d\tau = (g_m - g_n^*) \int \Gamma_n^* \Gamma_m \, d\tau \tag{6}$$

In general, the operators we will use have the property of being Hermitian. An operator is called Hermitian if the two terms in the left hand side of (6) are equal. Thus, if G is Hermitian

$$0 = (g_m - g_n^*) \int \Gamma_n^* \Gamma_m \, d\tau \tag{7}$$

When $m = n$, (7) is satisfied only if $g_m = g_m^*$; so the eigenvalues of a Hermitian operator must be real. For $m \neq n$, if the eigenvalues are distinct, that is, if $g_m \neq g_n$, then clearly (2) must be satisfied. If, however, Γ_m and Γ_n are degenerate eigenfunctions so that $g_m = g_n$, then (2) is not necessarily satisfied. We have seen in Chapter 3, however, that we are free to form linear combinations of degenerate eigenfunctions; thus, we can always choose Γ_m and Γ_n so that the orthogonality theorem (2) is satisfied, even in the event of degeneracy.

Suppose now that there are several operators whose eigenvalues we wish to find. Perhaps we have found the eigenfunctions (wavefunctions) and eigenvalues (associated energies) of the Hamiltonian for a system and wish to find the eigenvalues of an operator which will give us the total spin angular momentum and those of another which will tell us the behavior with respect to a symmetry element of each eigenfunction. Will the eigenfunctions of the Hamiltonian in general also be eigenfunctions of other operators? The answer is given by a theorem which says that two operators will have a simultaneous set of eigenfunctions if, and only if, the operators commute. This theorem will be most important to us for establishing that operators F and G which commute have a common set of eigenfunctions, so let us begin by assuming that F and G commute.

$$GF = FG \tag{8}$$

Operating with (8) on Γ_m, one of the eigenfunctions of G, we obtain

$$GF\Gamma_m = FG\Gamma_m = Fg_m\Gamma_m = g_mF\Gamma_m \tag{9}$$

Thus, $F\Gamma_m$ appears to be an eigenfunction of G with eigenvalue g_m. However, if Γ_m is nondegenerate, it is the only eigenfunction of G with eigenvalue g_m. Therefore, $F\Gamma_m$ must be a multiple of Γ_m. Thus,

$$F\Gamma_m = f\Gamma_m \tag{10}$$

where f is just a constant; so Γ_m is an eigenfunction of F as well as G. On the other hand, if Γ_m is degenerate, then there are other eigenfunctions of G with eigenvalue g_m. In this case, (9) is satisfied by

$$F\Gamma_m = \sum_{m'} f_{m'}\Gamma_{m'} \tag{11}$$

where the sum is over just those degenerate eigenfunctions with eigenvalue g_m. According to (11) the degenerate Γ_m will not in general be eigenfunctions of F, but will be transformed by F into linear combinations of each other. We found this to be the case for the initially chosen degenerate orbitals in cyclobutadiene, which were not eigenfunctions of the symmetry operations of reflecting through diagonal planes. Each of the two MO's was not transformed by these operations into a multiple of itself but rather into the other degenerate orbital. We were, however, able to make linear combinations of these orbitals that were eigenfunctions of these operators; and in general, we will be able to find linear combinations of degenerate eigenfunctions of one operator that will also be eigenfunctions of a second operator which commutes with the first.

Problem 6.1. Prove that if two operators have a simultaneous set of eigenfunctions (i.e. $F\Gamma_m = f_m\Gamma_m$ and $G\Gamma_m = g_m\Gamma_m$), the operators commute.

Problem 6.2. Suppose that a Hermitian operator F is defined such that $F\Gamma_1 = a_{11}\Gamma_1 + a_{12}\Gamma_2$ and $F\Gamma_2 = a_{21}\Gamma_1 + a_{22}\Gamma_2$, where Γ_1 and Γ_2 are orthonormal degenerate eigenfunctions of G. What constraints does Hermiticity place on the coefficients a_{12} and a_{21}? Show how to find two new functions which are eigenfunctions of F.

We have now shown that two commuting operators have a simultaneous set of eigenfunctions, so that we can classify each wavefunction for a molecule by its eigenvalue for any operator which commutes with the Hamiltonian. This is, of course, a convenient way of labelling wavefunctions; but we have seen that the real use of such labelling in the case of symmetry elements is that wavefunctions of different classifications are not mixed with each other by the Hamiltonian. We now show in general that if two eigenfunctions Γ_m and Γ_n of an operator F have different eigenvalues, the integral $\int \Gamma_n^* G\Gamma_m \, d\tau$ vanishes when G is an operator with which F commutes. Since we assume Γ_m is an eigenfunction of F,

$$\int \Gamma_n^* GF\Gamma_m \, d\tau = f_m \int \Gamma_n^* G\Gamma_m \, d\tau \tag{12}$$

Because F and G commute the left hand of (12) can be rewritten

$$\int \Gamma_n^* GF\Gamma_m \, d\tau = \int \Gamma_n^* FG\Gamma_m \, d\tau \tag{13}$$

We now assume that F is Hermitian so (13) becomes

$$\int \Gamma_n^* G F \Gamma_m \, d\tau = \int \Gamma_n^* F(G\Gamma_m) \, d\tau = \int (G\Gamma_m) F^* \Gamma_n^* \, d\tau$$

$$= f_n^* \int (G\Gamma_m)\Gamma_n^* \, d\tau = f_n^* \int \Gamma_n^* G \Gamma_m \, d\tau \qquad (14)$$

Subtracting (14) from (12) and using the fact that f_n is real,

$$0 = (f_m - f_n) \int \Gamma_n^* G \Gamma_m \, d\tau \qquad (15)$$

so that unless $f_n = f_m$, the integral vanishes. Thus, if we can find a number of operators which commute with the Hamiltonian, then, by classifying our wavefunctions with respect to them, we can quickly discover which wavefunctions will not be mixed by the Hamiltonian via an integral of the type in (15).

Some Operators Which Commute with the Hamiltonian

To a good approximation, because of the very small amount of spin-orbit coupling in first row elements, the Hamiltonian is spin free in the molecules we shall discuss. If there are no terms in the Hamiltonian involving spin, the Hamiltonian is invariant to all spin operators, so operators like S^2 and S_z will commute with it. Thus, not only will we be able to classify all our molecular states with respect to spin, but we will also be guaranteed that those with different spin quantum numbers will not mix, in so far as it is legitimate to neglect spin-orbit coupling. The usefulness of this fact will become apparent in Chapter 9. However, many of our problems are concerned with finding MO's as LCAO's and determining how MO's mix. For these tasks spin classifications are of no relevance. We have already found that use of molecular symmetry effects great simplifications in such problems, and we will now show more rigorously why molecular symmetry is of such utility.

As the Hamiltonian is invariant under a number of operations, a number of operators commute with it. For example, since electrons are indistinguishable, it does not matter which we arbitrarily designate as electron 1, 2, 3, etc. Therefore, the Hamiltonian is invariant to any permutation of the positions and spins of the electrons in a system, that is, to any operation that merely interchanges the positions and spins of the electrons that are arbitrarily called m and n. The Hamiltonian is also invariant to rotation about any axis through the center of mass and to any translation in space of the system. The fact that free space is homogeneous, so that there should be no preferred position or orientation of our molecular coordinate system, gives rise to these two invariancies. It can be shown that these invariancies are responsible for the laws of conservation of linear and angular momentum. The Hamiltonian is also unchanged by the reversal of all particle momenta and spins; in other words, it is invariant to time reversal.[1]

These invariancies of the Hamiltonian are very interesting; but, except for the electron permutation operator, to which we shall return shortly, the operators associated with translation and rotation of the molecular coordinate system in space and with time reversal are not very useful for classifying our electronic wavefunctions, since our wavefunctions are also invariant to all these operations. There are, however, two more operations to which the Hamiltonian is invariant that are useful for the classification of wavefunctions. One is the inversion of the positions of all particles through the center of mass, which arises from the equivalence of left- and right-handed coordinate systems and implies conservation of parity. Of course, in weak nuclear interactions it has been shown that parity is not conserved; but since we are discussing molecular and not nuclear physics, this need not trouble us. Finally, the Hamiltonian is invariant to any permutation of the positions and spins of identical nuclei. Clearly, we need not concern ourselves with all such possible permutations; since, for instance, the exchange of just the two carbon nuclei in ethane, but not the hydrogens attached to them, is not feasible on the time scale of most experiments. However, permutation of the three hydrogen nuclei in one of the methyl groups of ethane is possible because of the low barrier to rotation about the C—C single bond.

Longuet-Higgins[2] has shown that the combination of the operations of inversion through the center of mass and all feasible permutations of

---

[1] It can be demonstrated that the Hamiltonian for a system in an external magnetic field is not invariant to time reversal, unless one takes into account the fact that on time reversal, the current which is producing the magnetic field must also be reversed, thereby reversing the direction of the field.

[2] *Mol. Phys.* **6**, 445 (1963).

identical nuclei are equivalent, in rigid molecules, to the familiar operations of reflecting through symmetry planes and rotating about symmetry axes. However, he has demonstrated that in nonrigid molecules like ethane it is essential to make use of the permutation and inversion operators in order to obtain a complete description of the full symmetry present in such molecules. Since we will be confining ourselves to the discussion of rigid molecules, we will employ the more conventional planes and axes in describing their symmetry. We should keep in mind, however, that rotations and reflections are useful in describing the symmetry of a molecule, precisely because these symmetry elements permute identical nuclei. Because the Hamiltonian is invariant to such permutations, any such operation will commute with the Hamiltonian. Therefore, our wavefunctions must also be eigenfunctions of the operators which represent all the symmetry elements present in a molecule. Moreover, wavefunctions which belong to different eigenvalues of such operators (i.e. wavefunctions which have different symmetry behavior) will not be mixed by the Hamiltonian, a fact that we will continue to find very useful.

Group Theory

It can be shown that the symmetry elements present in a molecule have the following properties, when the identity operation, which permutes no nuclei, is included among them: (1) The effect of the successive application of two elements is identical to that produced by a single symmetry element in the set. (2) The identity element, usually denoted E, commutes with all the other symmetry elements and leaves them unchanged. (3) Although the commutative law of multiplication does not necessarily hold, the associative law does. Thus, if R, S, and T are operators which represent symmetry elements, RS is not necessarily equal to SR, but $(RS)T = R(ST)$. (4) Every element R has an inverse R^{-1}, which is also a symmetry element such that $RR^{-1} = E$. These four facts are sufficient to show that the symmetry elements present in a molecule must form a mathematical group; hence, molecular symmetry is amenable to treatment by the branch of mathematics known as group theory. It is not the purpose of this section to teach group theory; the interested reader is referred to one of the books suggested at the end of this chapter. However, in this limited space we will attempt to provide some rules through the use of which the reader will be able to apply the results of group theory to problems in which molecular symmetry exists.

Let us begin by considering a specific problem, that of a bent AB_2 molecule. In addition to E there are three other symmetry elements—a twofold axis and two orthogonal symmetry planes which contain the axis. These are shown in Figure 6.1.

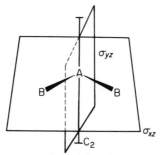

Symmetry Elements Present in a Bent AB_2 Molecule

Figure 6.1

Problem 6.3.   Show that the symmetry elements in Figure 6.1 form a group.

Having determined the symmetry elements present in the molecule, we must identify the group to which it belongs. This can be done by inspecting a list of the symmetry elements which comprise each of the different groups. The brief descriptions in Table 6.1 of the various types of symmetry groups may be helpful in assigning a molecule to the one to which it belongs.

Problem 6.4.   Choose five molecules drawn in the preceeding chapters and assign each to a symmetry group. To which symmetry group does allene, $CH_2{=}C{=}CH_2$, belong?

From Table 6.1 we find that our AB_2 molecule belongs to the group C_{2v}. With this information we can look up the *character table* of C_{2v}, which can be found in any one of the books suggested at the end of the chapter and which is reproduced in Table 6.2.

On the top line are listed the operations of the symmetry group. Across the four lines below it are listed four sets of numbers, called *characters*. In each set there is one character for each operation. The characters are really eigenvalues, one for each symmetry operator, and so they multi-

Table 6.1

Group	Symmetry Elements Present
C_n	One n-fold rotation axis
D_n	One n-fold axis
	n 2-fold axes perpendicular to it
C_{nv}	One n-fold axis
	n vertical (containing the axis) planes
C_{nh}	One n-fold axis, one horizontal plane, one improper n-fold axis*
D_{nd}	One n-fold axis, n 2-fold axes perpendicular to it, n vertical planes bisecting the angles between the 2-fold axes, one improper $2n$-fold axis* with only rotations of odd order allowed
D_{nh}	One n-fold axis, n 2-fold axes perpendicular to it, n vertical planes, one horizontal plane, one improper n-fold axis*
T_d	Tetrahedral symmetry
O_h	Octahedral symmetry
$C_{\infty v}$	Linear molecules
$D_{\infty h}$	Linear molecules with a plane perpendicular to the molecular axis

* An improper n-fold axis, S_n, requires a rotation of $360°/n$ and then a reflection through a perpendicular plane. Thus an S_2 axis, requiring rotation by 180° followed by reflection, is equivalent to an inversion center.

Table 6.2

C_{2v}	E	C_2	$\sigma(xz)$	$\sigma(yz)$		
A_1	1	1	1	1	z	x^2, y^2, z^2
A_2	1	1	-1	-1	R_z	xy
B_1	1	-1	1	-1	x, R_y	xz
B_2	1	-1	-1	1	y, R_x	yz

ply in the same way as the symmetry operations which they represent. Each set of characters is called a *representation* of the group, and the four sets of characters in the table represent all the possible effects of the symmetry elements on a wavefunction in a molecule of C_{2v} symmetry.

Problem 6.5. In Problem 6.3 you determined how the symmetry operations in C_{2v} multiply (e.g. $\sigma_{xy} \times \sigma_{xz} = C_2$). Show that each of the representations of C_{2v} follows the same multiplication rules. Why is the operation E always represented by the number 1?

To the left of each representation is a symbol which serves as a name for it. There is always a representation in which all the operations of a group are represented by $+1$. This is called the totally symmetric representation, since a function which belongs to it has eigenvalue $+1$ for all the operations of the group and so is unaffected by any of them. We can rapidly find out to which representation any wavefunction belongs by operating on it with the symmetry elements in the group. The set of eigenvalues so generated are just the numbers found in the character table for the representation to which it belongs. Moreover, we are guaranteed that wavefunctions belonging to different representations of the group will not be mixed by the Hamiltonian, since they will have different eigenvalues for some of the operations. Thus, once we have some wavefunctions for the AB_2 molecule, by finding the representation to which each belongs in C_{2v} symmetry, we can rapidly determine which will mix.

However, we have previously used symmetry to help us to find MO's. How can we use the formalism of group theory to do the same? Let us begin by noting from inspection of the character table that the different representations behave like orthogonal vectors; that is, when the characters of two different representations, i and j, for each operation R of the group are multiplied together and the products summed, the result is zero. If the same process is applied to identical representations (i.e. if a representation is "squared") the result is the number of operations, h, in the group. More concisely,

$$\sum_{R} \chi_j(R)^* \chi_i(R) = h\delta_{ij} \tag{16}$$

where $\chi_j(R)^*$ is the complex conjugate of $\chi_j(R)$, in the event that $\chi_j(R)$ has an imaginary part; and δ_{ij} is defined so that $\delta_{ij} = 1$ when $i = j$ and equals zero for $i \neq j$. An equation similar to (16) holds quite generally, and is very important in proving many theorems in the mathematics of groups. We will now employ (16) to show for nondegenerate representations how the character table may be used to determine the MO's of a system. The theorem also holds for degenerate representations, but the proof is a little more complicated.

Since for a system of n AO's we have n MO's written as LCAO's, by solving n equations for the n AO's it is possible to write the rth AO as a LCMO. Thus,

$$\phi_r = \sum_i c_{ri} \psi_i \qquad (17)$$

The result of operating on (17) with any group operation is

$$R\phi_r = \sum_i c_{ri} R\psi_i = \sum_i c_{ri} \chi_i(R) \psi_i \qquad (18)$$

where $\chi_i(R)$ is the eigenvalue of ψ_i for the operation R; or in the language of group theory, it is the character for the operation R of the representation to which ψ_i belongs. Suppose now that we wish to find the MO which belongs to the jth representation. If we take any AO, operate on it with R, multiply by $\chi_j(R)^*$, repeat this procedure for all operations of the group and sum, we obtain, using (18) and then (16),

$$\sum_R \chi_j(R)^* R\phi_r = \sum_R \sum_i c_{ri} \chi_j(R)^* \chi_i(R) \psi_i = \sum_i c_{ri} \psi_i h \, \delta_{ij} \qquad (19)$$

It is clear from (19) that if two MO's have the same symmetry, either by accident or because the jth representation is degenerate, the process indicated in (19) will give a linear combination of MO's of the same symmetry; but as we shall see shortly, this will pose no problem to finding the correct individual MO's.

Suppose now that we wish to find the MO's of a molecule which are formed from a set of symmetry equivalent AO's of which ϕ_r is one. We operate on ϕ_r with each symmetry element in the group, each of which will in general transform it into another AO. Multiplying the result of each operation by $\chi_j(R)^*$, the complex conjugate of the character of the jth representation for that operation, we obtain on summing a linear combination of MO's belonging to the jth representation. If there are no MO's, formed from the set of equivalent AO's of which ϕ_r is one, that belong to this representation, (19) informs us that we will obtain zero. Unless we can somehow determine in advance the representations spanned by the MO's formed from a set of AO's, then for each set of equivalent AO's we will have to repeat the process outlined in (19) for all the representations in the group. For a group containing many operations and many different possible representations, this can be a formidable task.

Therefore, we now turn our attention to finding a procedure for determining in advance which representations will be spanned by the MO's formed from a set of AO's. This is most simply done with the introduction of matrix notation.

Suppose that we have n AO's from which we wish to form n MO's. The MO's can be written as n equations of the form

$$\psi_j = \sum_s c_{js}\phi_s \tag{20}$$

For large n this number of equations becomes rather unwieldly. However, all n equations can be concisely represented in matrix notation as

$$[\psi_j] = [c_{js}][\phi_s] \tag{21}$$

where it is understood that the c's are written in a square array, or matrix, and that the ψ_j and ϕ_s are arrayed as column matrices. This means that we are defining matrix multiplication so that each element of the product matrix $[\psi_j]$ is given by the sum of products of each element in row j and column s of the matrix $[c_{js}]$ with the element in row s of $[\phi_s]$. More succinctly, the rule for matrix multiplication of $[X][Y] = [Z]$ may be stated as: "the elements z_{ik} of the product matrix are found by multiplying the ith row in X into the kth column of Y." In (22) the MO's for ethylene are expressed in matrix notation and multiplied out.

$$\begin{bmatrix} \psi_1 \\ \psi_2 \end{bmatrix} = \begin{bmatrix} \dfrac{1}{\sqrt{2}} & \dfrac{1}{\sqrt{2}} \\ \dfrac{1}{\sqrt{2}} & \dfrac{-1}{\sqrt{2}} \end{bmatrix} \begin{bmatrix} \phi_1 \\ \phi_2 \end{bmatrix} = \begin{bmatrix} \dfrac{1}{\sqrt{2}}\phi_1 + \dfrac{1}{\sqrt{2}}\phi_2 \\ \dfrac{1}{\sqrt{2}}\phi_1 - \dfrac{1}{\sqrt{2}}\phi_2 \end{bmatrix} \tag{22}$$

Problem 6.6. Write out explicitly the MO's of allyl in matrix notation as $[\psi_i] = [c_{ir}][\phi_r]$. Show that they can equivalently be written $[\psi_i] = [\phi_r][c_{ri}]$ where $[\psi_i]$ and $[\phi_r]$ are row matrices.

Problem 6.7. Write the n equations represented by (17) in matrix notation. Substitute for $[\phi_s]$ in (21) and thus demonstrate that the product matrix $[c_{ji}] = [c_{jr}][c_{ri}]$ has elements given by $c_{ji} = \sum_r c_{jr}c_{ri} = \delta_{ij}$. This is known as the unit matrix, since its only nonzero elements are 1's along the diagonal.

We have seen that the transformations between AO's and MO's can be concisely represented in matrix notation; and we shall see in subsequent chapters the general utility of matrices in quantum mechanical problems. Matrices are also useful in representing the effect of symmetry elements on a set of orbitals. For instance, if we operate on a set of MO's with a symmetry element R, we can express its effect with a matrix as in (23).

$$R[\psi_i] = [R_{ij}][\psi_j] \qquad (23)$$

Since all nondegenerate MO's are eigenfunctions of the symmetry operations of the group to which the molecule belongs, these MO's will either have a plus or minus 1 on the diagonal, depending on their eigenvalue for the operation. Degenerate MO's, however, may be transformed into each other by some symmetry operations and hence will have elements in the transformation matrix which are not on the diagonal.

Problem 6.8.   Write the results of Problem 3.9, the effect of the symmetry planes in cyclobutadiene on the primed and unprimed set of degenerate orbitals, in matrix notation.

Similarly, we can also represent the effect of a symmetry element on a set of AO's by a matrix equation like (23).

$$R[\phi_r] = [R_{rs}][\phi_s] \qquad (24)$$

In general, we may expect the matrices $[R_{ij}]$ and $[R_{rs}]$ to be very different. For example, most MO's will be transformed into ± 1 times themselves by R while most AO's will be transformed into other AO's. Thus, we anticipate that $[R_{ij}]$ will have most of its nonzero elements along the diagonal, while most of those in $[R_{rs}]$ will be off-diagonal. However, there is one very important feature that the two matrices have in common and which will allow us to find the representations spanned by the MO's directly from the AO's from which they are formed. We will now prove that both $[R_{ij}]$ and $[R_{rs}]$ have the same *trace*; that is, the sum of the diagonal elements of both is the same. Operating on the matrix form of (17) with R and substituting (23)

$$R[\phi_r] = R[c_{ri}][\psi_i] = [c_{ri}]R[\psi_i] = [c_{ri}][R_{ij}][\psi_j] \qquad (25)$$

The matrix $[c_{ri}]$ is just an array of constants, which is why it is unaffected by R in (25). Substituting in the left hand side of (25) from (24) and for $[\psi_j]$ in the right hand side from (21) we obtain

$$[R_{rs}][\phi_s] = [c_{ri}][R_{ij}][c_{js}][\phi_s] \qquad (26)$$

Thus,

$$[R_{rs}] = [c_{ri}][R_{ij}][c_{js}] \qquad (27)$$

The trace of $[R_{rs}]$ is just the sum of its diagonal elements, which is

$$Tr[R_{rs}] = \sum_r \sum_s R_{rs}\,\delta_{rs} = \sum_r \sum_s \left(\sum_i \sum_j c_{ri}R_{ij}c_{js}\right)\delta_{rs}$$
$$= \sum_i \sum_j R_{ij}\left(\sum_r c_{ri}c_{jr}\right) \qquad (28)$$

where we are now writing the actual matrix elements involved so that the quantities in (28) are just numbers. But in Problem 6.7 it was shown that

$$\sum_r c_{ri}c_{jr} = \delta_{ij} \qquad (29)$$

so, using (29) we obtain

$$\sum_r \sum_s R_{rs}\,\delta_{rs} = \sum_i \sum_j R_{ij}\,\delta_{ij} \qquad (30)$$

Thus, we have shown that $[R_{ij}]$ and $[R_{rs}]$, or any two matrices related by the transformation (27), have identical traces when the transforming matrices have elements which satisfy (29).

Obtaining the trace of $[R_{rs}]$ is trivial. It is just the number of equivalent AO's that are left unchanged by a symmetry operation, minus the number of AO's that are transformed into the negative of themselves. We have shown that the trace of $[R_{rs}]$ is the same as that of $[R_{ij}]$, and the trace of $[R_{ij}]$ is just the sum of the characters (eigenvalues) of all the MO's on the operation R. Thus, from the above process for finding the trace of $[R_{rs}]$ for the operation R, we also obtain for that operation the sum of the characters of all the MO's formed from the AO's. Such a sum of characters is obtained for each operation of the group; and by inspection of the character table we can then, by trial and error, deduce the representations whose characters must be added to give the same sum.[3] The representations spanned by the MO's, formed from a set of AO's, can thus be determined from the transformation properties of the AO's.

The problem of finding the MO's in our bent AB_2 molecule, because of its simplicity, does not demonstrate the full power of group theoretical methods; however, the simplicity of this problem makes it useful for illustrating clearly how group theory is actually applied. We begin by finding the representations of the MO's formed from the hybrid σ bonding orbitals on the two B atoms. Each hybrid, of course, is transformed into itself by E, the identity operation; so the character for this operation is two. Each is also transformed into itself on reflection through σ_{xz}. However, the orbitals are each transformed into the other by the second plane of symmetry and the twofold axis; so the character for these last operations is zero. These numbers, shown in the following table, are also the sums of the characters, for each operation in the group, of the MO's formed from the hybrids.

E	C_2	$\sigma(xz)$	$\sigma(yz)$
2	0	2	0

It is easy to see from the C_{2v} character table that the two representations which give this sum are A_1 and B_1; and we now proceed to find the actual MO's which belong to these representations. If we begin with ϕ_1, the hybrid on atom 1, the results of operating on this orbital with all the operations of the group are the following.

E	C_2	$\sigma(xz)$	$\sigma(yz)$
ϕ_1	ϕ_2	ϕ_1	ϕ_2

To actually find the MO's we need, for each operation, to multiply the transformed ϕ_1 by the character of the representation to which the MO

---

[3] Consult one of the texts listed at the end of the chapter for a more systematic procedure for determining from which representations such a sum of characters arises. The procedure utilizes (16) and is thus rather tedious in that it requires us to multiply the sum by each representation in order to determine whether it occurs; and if so, how many times. For most problems in MO theory the trial and error approach usually proves quickest.

belongs and then to sum the results. Thus, we find the two unnormalized MO's to be

$$\psi_{A_1} = \phi_1 + \phi_2 + \phi_1 + \phi_2 \quad \text{and} \quad \psi_{B_1} = \phi_1 - \phi_2 + \phi_1 - \phi_2$$

Problem 6.9. Show mathematically that the two hybrids, ϕ_1 and ϕ_2, do not form a basis for MO's of A_2 and B_2 symmetry. Explain why.

Clearly, these are not really MO's; they are only the parts of MO's that involve the two equivalent bonding hybrid orbitals on the terminal atoms. We can think of each of these MO fragments that we have found as arising from a linear combination of the MO's of its symmetry, such that all the contributions to each fragment from other AO's have been cancelled out. To reconstruct the complete MO's of each symmetry, we must find the other AO's in the molecule that belong to the A_1 and B_1 representations and which will consequently mix with the combination of hybrid orbitals of the same symmetry on the terminal atoms.

Because the $2s$ AO on the central atom has spherical symmetry, it is clearly invariant to all the operations of the group and must belong to the A_1 representation. The representations to which the p orbitals belong can be found directly from the third area of the character table where rotations about the three coordinate axes and vectors along the axes are grouped according to which representation each belongs. Since each p orbital has the symmetry behavior of a vector along one of the coordinate axes, we can immediately see that the p_z orbital on the central atom has A_1 symmetry, the p_x B_1, and the p_y B_2 symmetry. These AO's mix with the MO fragments of the same symmetry on the terminal atoms to form the actual MO's for bent AB_2.

Problem 6.10. Find the symmetry of the MO fragments formed from the remaining sets of p orbitals on the B atoms. Find these MO fragments. Make a table showing the AO's or combinations of AO's of the same symmetry in a bent AB_2 molecule. Pick out the AO's that formed the MO's you used in Problem 2.20. Did you miss any interactions between MO's in working Problem 2.20?

When a molecule contains a symmetry axis greater than twofold (i.e. a C_n or S_n axis with $n > 2$), some of the representations in the character table for the group to which it belongs are degenerate. This means that functions which belong to these representations will be transformed into each other by some of the symmetry operations of the group. The entries in the character table for a degenerate representation cannot, therefore, be the eigenvalues of such functions for every operation, since a function that is transformed into another by an operator is not an eigenfunction of that operator. Instead, the characters of a degenerate representation are the trace of the matrix which represents the effect of each operation on functions which belong to that representation. If each of two[4] degenerate functions is transformed into itself, for instance, by the identity operation, then the character for that operation is two. If both are transformed into each other, the character is zero, and so on.

Let us actually find the MO's for a molecule which belongs to a group containing degenerate representations. A planar CH_3 fragment belongs to the group D_{3h}, whose character table appears as Table 6.3.

Table 6.3

D_{3h}	E	$2C_3$	$3C_2$	σ_h	$2S_3$	$3\sigma_v$		
A_1'	1	1	1	1	1	1		$(x^2 + y^2, z^2)$
A_2'	1	1	-1	1	1	-1	R_z	
E'	2	-1	0	2	-1	0	(x, y)	$(x^2 - y^2, xy)$
A_1''	1	1	1	-1	-1	-1		
A_2''	1	1	-1	-1	-1	1	z	
E''	2	-1	0	-2	1	0	(R_x, R_y)	(xz, yz)

There are several points worthy of note in this character table. First, some operations are grouped together; for instance, rotations of $120°$ and $240°$ about the threefold axis are listed together under $2C_3$. Similarly, the three twofold axes, each of which runs through a C—H bond in planar CH_3, are collectively referred to in the character table as $3C_2$. The reason that certain operations are listed together is that they have the same effect on functions in molecules of D_{3h} symmetry; therefore, in each representation they have the same character. Such operations are said to belong to the same *class*, and we must be careful to remember

---

[4] Only in molecules of tetrahedral or octahedral symmetry does one encounter representations with degeneracy greater than twofold. Threefold degenerate representations are given the label T, twofold E, and nondegenerate representations, as we have seen, are labelled A or B.

that a class may contain several distinct operations. This is especially important in generating MO's, since (19) requires that we operate on an AO with every operation in the group, not just one from each class.

Another item of interest in D_{3h} is that the second group of three symmetry operations in the table can be obtained from the first three by multiplying each by σ_h, the operation of reflection through the horizontal symmetry plane. The first three operations together form the symmetry group D_3; addition of the horizontal plane gives D_{3h}. The former is said to be a *subgroup* of the latter. Note also that the representations of D_{3h} are closely related to those of D_3. Functions which are symmetric to the horizontal symmetry plane have for the last three operations of D_{3h} characters that are identical to those for the operations which comprise the D_3 subgroup. Such functions belong to singly primed representations, while functions which belong to doubly primed representations are antisymmetric to the plane and, consequently, have characters for the last three operations that are the negative of those for the first three.

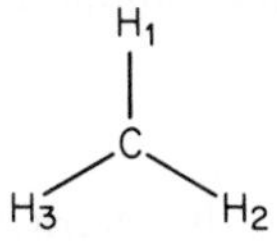

Planar Methyl

Figure 6.2

Let us now use the D_{3h} character table to find the MO's for planar CH_3. We begin by finding the symmetry of the MO's that will be formed from the three equivalent hydrogen $1s$ orbitals. We could use all the operations of D_{3h} to accomplish this task; however, we can be sure that, because these AO's are symmetric to σ_h, only singly primed MO's can be formed from them. Therefore, we can restrict our attention to determining how these three hydrogen $1s$ AO's transform under the operations of just the D_3 subgroup. The identity operation leaves all three orbitals alone; obviously, the character for this operation is then 3. A rotation of the molecule by 120° or 240° takes each hydrogen into a different one; the character for this class of operations is zero. Finally, a rotation about any of the three C_2 axes interchanges two hydrogens but leaves alone the one through which the axis passes; the character for this class is 1. The representations spanned by the three hydrogen AO's have characters that sum to 3, 0 and 1 for the three classes of the D_3 subgroup. These characters can be reduced to the sum of those for A_1 and E, so we know the three hydrogen atoms will form MO's of A_1' and E' symmetry. We can also find the MO fragments of each symmetry, formed by the hydrogen atoms, through use of the character table and Equation (16). Starting with hydrogen atom 1, we find that the identity operation leaves it alone, one of the rotations about the threefold axis takes it into atom 2, and the other into 3. One of the three C_2 axes leaves H_1 alone, one interchanges it and 2, and the last switches it and 3. A summary of the results follows.

E	$2C_3$	$3C_2$
H_1	H_2, H_3	H_1, H_2, H_3

The character of the A_1 representation is 1 for each operation. On summing we obtain the unnormalized $\psi_{A_1'} = H_1 + H_2 + H_3$. The $2s$ AO on the central carbon atom is also invariant to all the operations of the group, so it also has A_1' symmetry and will mix with the above combination of hydrogen atoms of this symmetry to form a bonding and an antibonding MO.

Problem 6.11. According to the character table the p_x and p_y orbitals on the central carbon atom have the proper symmetry for mixing with the E' combination of hydrogen atoms. Show that together these two AO's do, in fact, have E' symmetry by determining the trace of the 2×2 transformation matrix for these orbitals for the operations of D_{3h}.

The combinations of hydrogen $1s$ orbitals belonging to the degenerate E' representation require a little more work to find. One of them is quite straightforward. Using the sequence of AO's that we generated starting with H_1, multiplying by the character of E for each operation, and summing, we obtain $2H_1 - H_2 - H_3$. Another sequence can be generated starting with H_2. Repeating the above process we obtain from it $2H_2 - H_3 - H_1$. This is not orthogonal to the first; and neither is $2H_3 - H_1 - H_2$, which can be obtained by starting with H_3. However, we are free to take linear combinations of degenerate MO's, and we can thus obtain an orbital orthogonal to the first one from the last two. Summing the last two gives the negative of the first degenerate orbital we found, but subtracting them gives $H_2 - H_3$, which is orthogonal to the first. We will find that in dealing with degenerate representations we often will not be able to generate two degenerate orthogonal orbitals directly from the character table but that we will have to create the second one from a linear combination of orbitals generated from the character table. As we shall have occasion to see in the next chapter, two degenerate orbitals that are orthogonal can also be found in complex form directly from the character table for special subgroups of the molecular symmetry group.

Problem 6.12.   How would you orient the p_x and p_y AO's on the central carbon atom to make the remaining problem of calculating the interaction between these orbitals and the degenerate E′ combinations of hydrogen orbitals as easy as possible to solve? Write general expressions for the three σ bonding symmetry MO's of planar CH_3 in terms of undetermined mixing coefficients between the hydrogen orbitals and those on the central carbon atom. Show that it is possible to obtain localized C—H bonding orbitals from linear combinations of these delocalized symmetry orbitals. Show that despite the fact that the central carbon is trigonal, symmetry does not require the orbitals it uses to form these localized bonds to be sp^2 hybrids. [Hint: Show that only if the mixing coefficients of the 2s orbital in the A′$_1$ MO and the 2p in the E′ are identical will the contribution of carbon to each of the localized orbitals be one third s and two thirds p. State in words why the mixing coefficients need not be the same.]

Problem 6.13.   Trimethylenemethane consists of a central carbon bonded to three CH_2 groups so that the π system of the neutral molecule contains four electrons. Dowd has synthesized this molecule, and it appears to have a triplet ground state. Without performing a Hückel calculation, use group theory to rationalize this observation.

Antisymmetrization of Electronic Wavefunctions

We have now seen some of the useful consequences of the fact that any operator which carries out a feasible permutation of identical nuclei commutes with the Hamiltonian. However, we previously have noted that any permutation of the coordinates and spins of electrons also leaves the Hamiltonian invariant; therefore, an electronic wavefunction must also be an eigenfunction of an operator which carries out such a permutation. Let us define one such operator, P_{ij}, which permutes the coordinates and spins of the ith and jth electrons, by the equation

$$P_{ij}\Psi(\ldots, i, \ldots, j, \ldots) = \Psi(\ldots, j, \ldots, i, \ldots) \qquad (31)$$

The operator in (31) has been written as permuting the labels of the ith and jth electrons in the wavefunction, which is equivalent to permuting their coordinates and spins. Since Ψ must be an eigenfunction of P_{ij}, we can also write

$$P_{ij}\Psi(\ldots, i, \ldots, j, \ldots) = \lambda \Psi(\ldots, i, \ldots, j, \ldots) \qquad (32)$$

What is the eigenvalue λ? We can find out be applying P_{ij} once again to (31) and (32) and equating the results.

$$\begin{aligned} P_{ij}\Psi(\ldots, j, \ldots, i, \ldots) &= \Psi(\ldots, i, \ldots, j, \ldots) \\ &= \lambda^2 \Psi(\ldots, i, \ldots, j, \ldots) \end{aligned} \qquad (33)$$

Equation (33) shows that $\lambda^2 = 1$, so $\lambda = \pm 1$. It is found experimentally that electrons and other particles of half-unit spin obey Fermi-Dirac statistics, which implies that they have $\lambda = -1$.

Problem 6.14.   Does the wavefunction $\Psi(1, 2) = \Phi_1(1) \ \Phi_2(2)$, where Φ_1 and Φ_2 are different one-electron wavefunctions, satisfy (32) with $\lambda = -1$? Construct a wavefunction for one electron in Φ_1 and another in Φ_2 that satisfies (32). Show that the requirement that electronic wavefunctions satisfy (32) with $\lambda = -1$ implies the more familiar statement of the Pauli exclusion principle, that two electrons cannot have the same wavefunction.

Problem 6.15.   The one-electron wavefunctions, Φ_1 and Φ_2, in the preceding problem must be products of both space and spin functions for each electron. Let ψ_1 and ψ_2 be two different spatial wavefunctions and α and β the two different possible spin functions for an electron. Write a correct wavefunction corresponding to two electrons occupying the same orbital with different spins. Write another where the electrons are in different orbitals with the same spin. How many other wavefunctions can you write for two electrons in different orbitals which satisfy (32)?

Since we require that our wavefunctions be antisymmetric to the permutation of the labels of any two electrons, and because any other permutation can be expressed as a product of two-electron permutations, a more general condition for a many-electron wavefunction is

$$P_\nu \Psi = (-1)^{m_\nu} \Psi \tag{34}$$

where m_ν is the number of two-electron permutations which effects the overall permutation P_ν. It can be shown that (34) can be satisfied by a wavefunction of the form

$$\Psi = \frac{1}{\sqrt{n!}} \sum_\mu (-1)^{m_\mu} P_\mu \Phi_1(1)\Phi_2(2)\ldots\Phi_n(n) \tag{35}$$

where the sum is over all the possible permutations of electrons. For example, a correctly antisymmetrized three-electron wavefunction is shown in (36).

$$\begin{aligned}
\Psi(1, 2, 3) = \frac{1}{\sqrt{6}}[&\Phi_1(1)\Phi_2(2)\Phi_3(3) - \Phi_1(1)\Phi_3(2)\Phi_2(3)\\
+ &\Phi_2(1)\Phi_3(2)\Phi_1(3) - \Phi_3(1)\Phi_2(2)\Phi_1(3)\\
+ &\Phi_3(1)\Phi_1(2)\Phi_2(3) - \Phi_2(1)\Phi_1(2)\Phi_3(3)]
\end{aligned} \tag{36}$$

Problem 6.16. Show that the wavefunction in (36) is normalized if the Φ_i are orthonormal.

As the reader is probably aware, the number of permutations of n objects is $n!$; so for the π system in benzene the number of terms in the wavefunction for the ground state of the molecule will be 720. Clearly, we will want to avoid actually writing out this wavefunction term by term, as we have done for three electrons in (36). Therefore, we note that the summation in (35) is just the definition of the determinant

$$\Psi = \frac{1}{\sqrt{n!}} \begin{vmatrix} \Phi_1(1) & \Phi_1(2) & \cdots & \Phi_1(n) \\ \Phi_2(1) & \Phi_2(2) & \cdots & \Phi_2(n) \\ \vdots & & & \\ \vdots & & & \\ \Phi_n(1) & \Phi_n(2) & \cdots & \Phi_n(n) \end{vmatrix} \tag{37}$$

Thus, we can avoid actually having to write out all the terms by writing the wavefunction as a determinant, as suggested by Slater. But we can do even better than this. We note that the diagonal contains all the information necessary to define the determinant; therefore, we will subsequently write our many-electron wavefunction as kets (the right-hand half of brac-kets)

$$\Psi = |\Phi_1(1)\Phi_2(2)\cdots\Phi_n(n)\rangle \equiv |\Phi_1\Phi_2\cdots\Phi_n\rangle \qquad (38)$$

where it is understood that our kets are really abbreviations for Slater determinants with the normalization factor of $1/\sqrt{n!}$ included. We will often, as indicated in (38), wish to drop the electron labels in parentheses; and in so doing, it is understood that the first one-electron wavefunction belongs to electron one, the second to two, etc.

Problem 6.17. Show by expansion of the Slater determinant that the wavefunction $\Psi = |\Phi_1\Phi_2\Phi_3\rangle$ is identical to (36). [Hint (for those unfamiliar with determinants): To expand a determinant one begins with the element in the first row and first column (a_{11}) and multiplies it times its minor (A_{11}), which consists of all the elements in the array *except* those in the first row or first column. The next term in the expansion is $-a_{12}A_{12}$, where a_{12} is the element in the first row and second column, and A_{12} is the minor formed from all the elements in the array except those in the first row or second column. The determinant can also be expanded along the first column, or any other row or column; but assuming we expand along the first row we finally get

$$\text{Det}|A| = a_{11}A_{11} - a_{12}A_{12} + a_{13}A_{13} - a_{14}A_{14} + \cdots(\pm)a_{1n}A_{1n} \qquad (39)$$

where it should be clear that those terms, the sum of whose indices is odd, occur with a minus sign in the summation. Each of the original minors (A_{1i}) is now expanded like the original determinant until the expansion by minors is complete.]

Because our many-electron wavefunctions behave like determinants, we can deduce several useful facts about them from the properties of determinants. For instance, if two rows or two columns are identical, a determinant vanishes. Since each row of a Slater determinant is a one-electron wavefunction, we see immediately that our many-electron wavefunctions, when written as Slater determinants, will obey the Pauli exclusion principle. Another useful property of determinants is that interchange of two rows or two columns changes the sign. Suppose our wavefunction is $\Psi = |\Phi_1\Phi_2\Phi_3\Phi_4\rangle$, and we wish to determine whether the term $\Phi_3\Phi_2\Phi_1\Phi_4$ occurs with a plus or minus sign in the expansion of the determinant, without actually carrying out the tedious expansion. If our wavefunction were $|\Phi_3\Phi_2\Phi_1\Phi_4\rangle$, the term $\Phi_3\Phi_2\Phi_1\Phi_4$ would be made up of all the diagonal elements in the determinant and would occur in the expansion with a plus sign. Since this second Slater determinant is related to the original by an odd number of row changes, the term of interest must appear with a minus sign in Ψ. We can use one more property of determinants—that any multiple of a row or column can be added to any other row or column without changing the value of the determinant —to prove a statement made in Chapter 2. The contention was made that if two orbitals are mixed by the Hamiltonian, when the two resulting orbitals are equally occupied, the situation with respect to the total energy is the same as if the electrons were localized in the original orbitals. In Chapter 2 we saw that energetically this was reasonable; now we demonstrate it to be true.[5] Suppose our wavefunction is

$$\Psi = |\ldots\Phi_i\ldots\Phi_j\ldots\rangle \tag{40}$$

where Φ_i and Φ_j are defined

$$\Phi_i = c_1\psi_1^\alpha + c_2\psi_2^\alpha \quad \text{and} \quad \Phi_j = c_2\psi_1^\alpha - c_1\psi_2^\alpha \tag{41}$$

so that they are orthogonal; and we assume $c_1^2 + c_2^2 = 1$, so that they are normalized. If we now add

$$\frac{c_2}{c_1}\Phi_j = \frac{c_2^2}{c_1}\psi_1^\alpha - c_2\psi_2^\alpha \tag{42}$$

to each element in the ith row, we do not change the value of the Slater determinant, which can be written

$$\Psi = \left|\ldots\left(c_1 + \frac{c_2^2}{c_1}\right)\psi_1^\alpha\ldots\Phi_j\right\rangle = \frac{c_1^2 + c_2^2}{c_1}\left|\ldots\psi_1^\alpha\ldots\Phi_j\right\rangle \tag{43}$$

[5] Actually, the perceptive reader will note that the following proof depends on the overlap integral between the orbitals being zero or negligibly small. The energetic argument given in Chapter 2 also depends on S being small. To the extent that S is not negligible, the net energy of the molecule is raised by the interaction between the orbitals, since a nonzero S leads to a bonding orbital which is less stabilized, compared to the starting orbitals, than the antibonding orbital is destabilized.

the last equality coming from the fact that multiplying every element in a row by the same constant is equivalent to multiplying the entire determinant by that constant. Now adding $-c_2 \, \psi_1^\alpha$ to the jth row we obtain

$$\Psi = \frac{c_1^2 + c_2^2}{c_1} \Big| \ldots \psi_1^\alpha \ldots -c_1\psi_2^\alpha \ldots \Big\rangle = -(c_1^2 + c_2^2)| \ldots \psi_1^\alpha \ldots \psi_2^\alpha \ldots \rangle$$

$$= | \ldots \psi_2^\alpha \ldots \psi_1^\alpha \ldots \rangle$$

(44)

thus proving our theorem. The reader should note that the theorem does not apply when the electrons have opposite spins.

The addition and subtraction of rows of a Slater determinant allow us to mix MO's without changing the overall many-electron wavefunction. Of course, the new MO's obtained by such mixing may not be eigenfunctions of any of the symmetry operators or of a one-electron energy operator, but these new MO's may have other desirable properties which make them useful. For instance, the MO's which are eigenfunctions of the symmetry operators are generally delocalized over the entire molecule; and for the organic chemist accustomed to thinking in terms of localized bonds, it is often convenient to transform to a set of localized orbitals, as in Problem 6.12, by mixing the delocalized symmetry orbitals. However, when a molecular state truly is delocalized, so that no single classical resonance structure can adequately represent it, then, of course, no amount of mixing of the delocalized orbitals will give a set of orbitals that is fully localized between just pairs of atoms.

Problem 6.18.   Consider the two equivalent double bonds in norbornadiene. Call the two bonding ethylene MO's π_1 and π_2. To what symmetry group does the molecule belong? Find the MO's which are eigenfunctions of the symmetry operators in this group and classify each according to the representation to which it belongs. Write the ground state wavefunction of the molecule as a Slater determinant, using the symmetry orbitals. Show that this is equivalent to the more familiar formulation of the ground state wavefunction, $\Psi = |\pi_1^\alpha\pi_1^\beta\pi_2^\alpha\pi_2^\beta\rangle$. Despite the fact that the localized and delocalized orbitals give the same many-electron wavefunction, and hence, total energy, will the first ionization potential of norbornadiene necessarily be the same as that of a mono-olefin?

Problem 6.19.   Show by trying to mix the two lowest MO's in butadiene that the ground state of this molecule cannot be adequately represented by two localized ethylene bonding MO's.

Separation of Electronic and Nuclear Wavefunctions—Born-Oppenheimer Approximation

We have discussed the electronic wavefunction and its behavior with respect to electron permutation operators, and we have discussed nuclear permutation operators. Nowhere, however, have we even mentioned the wavefunction for the nuclei. The reason for this omission is that, to a good approximation, the electronic wavefunction may be found for a given set of nuclear coordinates by assuming that the nuclei are fixed at those positions. This approximation, known as the Born-Oppenheimer approximation, works well because electrons, being less than one-thousandth the mass of even a hydrogen nucleus, move much faster than nuclei. Thus, to the rapidly moving electrons the nuclei appear virtually fixed. On the other hand, to the slow moving nuclei the positions of the electrons at any instant in time is of little importance, since the nuclear motion is so slow that the nuclei can only respond to the average electronic distribution. Therefore, the total wavefunction may be written as a product of electronic and nuclear parts

$$\Psi(r, R) = \Psi_e(r, R)\Psi_N(R) \qquad (45)$$

where the electronic wavefunction depends on the instantaneous coordinates of both the electrons and nuclei, while the nuclear wavefunction depends only on those of the nuclei. The two wavefunctions may be found separately by solving two different Schrödinger equations. First, the Schrödinger equation for the electrons is solved by assuming, as we have done throughout, that the nuclei are fixed at a given set of coordinates. The solution of a second Schrödinger equation for the nuclear wavefunction, in which the average electron distribution provides part of the potential in which the nuclei move, may then be undertaken. Thus, within the limits of validity of the Born-Oppenheimer approximation,[6] if we

---

[6] The Born-Oppenheimer approximation breaks down if the electronic wavefunction becomes a very sensitive function of the nuclear coordinates, so that it is strongly coupled to the nuclear wavefunction. This occurs near a point where two electronic states have the same energy, for instance, in a molecule predicted to undergo Jahn-Teller distortion. In such a situation the electronic wavefunction changes rapidly with small changes in the nuclear coordinates.

are solely interested in the behavior of the electrons in molecules, we may ignore the nuclear wavefunction.

Vibrationally Induced Mixing of Electronic Wavefunctions—Use of the Symmetry of Molecular Vibrations

The separability of the nuclear and electronic wavefunctions does not mean that the wavefunctions for nuclei have no effect on processes involving electrons. For instance, the shape of bands in electronic spectra is greatly influenced by the relation between the nuclear vibrational wavefunctions in ground and excited states. Nor should the fact that the Born-Oppenheimer approximation allows us to ignore the nuclear wavefunctions in doing calculations on the electrons be construed to mean that they cannot be utilized profitably in problems involving the electrons. In particular, great use can be made of the fact that, like the electronic wavefunctions, the nuclear wavefunctions can be classified with respect to their behavior to symmetry operators.

Suppose, for instance, that we have calculated the MO's for a molecule in its most symmetrical configuration. Since the molecule is vibrating, we may wish to see how a vibration that takes the molecule into a less symmetrical configuration changes the MO's. Now although we might just recalculate the MO's for this new configuration, we saw in dealing with triatomics in Chapter 2 that the application of perturbation theory to this problem is often judicious. In Chapter 2 we approached the problem of finding the MO's for the new molecular geometry by reclassifying the old MO's, according to which representation each belonged in the new molecular symmetry group, to see which would mix. However, another approach is possible. Since the perturbation to the Hamiltonian caused by the change in nuclear coordinates is brought about by a molecular vibration, the perturbation has the same symmetry as the vibration. The perturbation, provided that it is not totally symmetric so that the molecular symmetry remains unaltered, will, we know, result in the mixing of MO's. Instead of assigning the old MO's to the representations of the new symmetry group in order to determine which will mix, we can determine this directly within the *original* group by using the symmetry of the molecular vibration that is responsible for the perturbation.

The mixing coefficient between two wavefunctions, ψ_i and ψ_j, is given in second order perturbation theory by Equation (22) of Chapter 2 as

$$c_{ij} = \frac{\int \psi_{i0} \mathcal{3C}' \psi_{j0}}{E_{i0} - E_{j0}}$$

where $\mathcal{3C}'$ is the perturbation to the Hamiltonian. In order for mixing to occur, it is obviously necessary that the integral in the numerator be different from zero. In general, an integral will be nonzero if, and only if, the integrand belongs to the totally symmetric representation of the symmetry group. This important rule is an extension of the familiar one (mentioned in Chapter 5) that only symmetric functions give nonzero integrals. The more general rule may be justified by the following argument. Suppose an integrand does not belong to the totally symmetric representation. Then some operation of the group will take the integrand, and hence the integral, into its negative. However, integrals are just numbers and cannot be altered by reflections, rotations, etc. The only number that can be taken into its negative and yet remain unchanged is zero; therefore, integrals whose integrands do not belong to the totally symmetric representation are zero. We can determine the representation to which an integrand belongs by taking the *direct product* of the representations of the functions in it. The direct product is obtained by multiplying together the characters of the representations for each operation of the group and deducing which representation(s) the new set of characters, thus obtained, span.

Problem 6.20.   In the group D_{3h} verify the following direct products: $A_2' \times A_2'' = A_1''$, $A_2' \times E' = E'$, and $E'' \times E'' = A_1' + A_2' + E'$. Why, when two singly or two doubly primed representations are multiplied, does one obtain a singly primed representation; while, when a singly and doubly primed representation are multiplied, a doubly primed representation results?

Two rules about direct products are easily verified: (1) The direct product of the totally symmetric representation with any other belongs to (has the symmetry of) the latter. (2) Only the direct product of a representation with itself gives (or in the case of a degenerate representation, contains) the totally symmetric representation. The second rule implies that when the product of two functions is integrated, the integral is zero unless the functions belong to the same representation.[7] It also follows that in an integral of the type $\int \psi_i G \psi_j \, d\tau$, where G is some operator, the direct product of the representations to which ψ_i and ψ_j belong must span the representation to which G belongs, or the integral will be zero. Our by-now-familiar adage that only wavefunctions of the same symmetry are mixed by the Hamiltonian can be seen as a specific case of this rule. Since the Hamiltonian is invariant to all the operations of a group, it must belong to the totally symmetric representation. Therefore, two wave-functions will only be mixed by the Hamiltonian if their direct product also spans the totally symmetric representation, which requires that they must belong to the same representation.

However, not all operators necessarily belong to the totally symmetric representation. An example of one that does not is the dipole operator, $e\mathbf{r}$, which determines whether an electronic transition is allowed. Another is a perturbation due to a change in molecular symmetry, which belongs to the same representation in the symmetry group for the molecule as the vibration that produces the perturbation. It is necessary that for such a perturbation to result in the mixing of two wavefunctions, the direct product of the representations to which the wavefunctions belong must span the representation to which the vibration belongs. We can use this fact to find the type of vibration which results in the mixing of two given wavefunctions. For instance, in benzene there are two low energy "forbidden" electronic transitions which, nevertheless, occur with appreciable intensity. A mechanism through which each can gain "allowed-ness" is by mixing into itself, under the influence of a vibration of proper symmetry, some of the wavefunction for the excited state to which a transition is allowed. Knowing the representations to which the excited states of the forbidden and allowed transitions belong, one can deduce the type of vibration required to mix them (See Problem 7.19).

In a molecule like benzene the largest mixing, caused by the per-turbation due to a non-totally symmetric vibration, is between excited states, because of their energetic proximity. However, mixing of the ground state of a molecule with one or more excited states is important in the application of vibronically induced orbital mixing to the prediction

---

[7] Even if two functions belong to the same representation, they may, of course, still be orthogonal. An example is provided by two different eigenfunctions of the Hamil-tonian that have the same symmetry.

of the modes by which molecules will tend most easily to distort. Molecular distortion will be easiest in a vibrational mode which results in the mixing of a filled orbital with one or more unfilled orbitals. In order that the mixing be large, the orbitals should be as close as possible in energy. Therefore, a molecule with high-lying filled and low-lying empty orbitals is a prime candidate for undergoing a "pseudo" or "second order" Jahn-Teller distortion. A reference is given at the end of this chapter to a paper in which Walsh's rules are rederived using this formalism. A fuller discussion of vibronically induced mixing of electronic wavefunctions cannot be included here, since to go deeply into the symmetry of molecular vibrations at this point would take us too far afield. However, an example of the vibronic mixing approach to Walsh's rule is given in the following problem, and the reader interested in learning more about the symmetry properties of molecular vibrations is referred to one of the texts on group theory listed at the end of the chapter.

Problem 6.21. The symmetry of the lowest MO's in planar methyl is A'_1, E', A''_2, the first two being C—H bonding combinations, and the third the nonbonding carbon p orbital. Because $2s$ orbitals are lower in energy than $2p$ orbitals, the A'_1 MO lies lower than the E'; and the out-of-phase combination of $2s$ and the hydrogen orbitals, also of A'_1 symmetry, is the lowest energy antibonding MO. A vibration of A''_2 symmetry is required to convert planar to pyramidal methyl. Use the character table of D_{3h} to show that the methyl cation is likely to remain planar but the radical might show some tendency to become pyramidal and the anion an even greater tendency.

Summary

Hückel theory has a number of shortcomings that arise from its failure to include electron repulsion explicitly in the Hamiltonian. In this chapter we laid the groundwork for the explicit treatment of electron repulsion by examining the form required of our wavefunctions. We began by proving a number of theorems regarding Hermitian operators. In particular, we demonstrated that they have real eigenvalues and that their eigenfunctions, when the eigenfunctions belong to different eigenvalues, are orthogonal. We also proved that the eigenfunctions of a Hermitian operator like the Hamiltonian must also be eigenfunctions of any operator which commutes with it; moreover, we showed that eigenfunctions belonging to different eigenvalues of an operator which commutes with the Hamiltonian are not mixed by the Hamiltonian.

This useful theorem led quite naturally to a discussion of group theory; for the operators which represent the symmetry elements in a symmetry group commute with the Hamiltonian, because of the invariancy of the Hamiltonian to the permutation of identical nuclei. We showed that by using the character table of the symmetry group to which a molecule belongs, it is possible to find the MO's or MO fragments formed from a set of AO's by the following procedure: First, one finds the signed total of AO's that are transformed into themselves or into their negatives by each operation of the group. This gives for each operation the sum of the characters of the representations spanned by the MO's formed from the AO's; and from this sum, the representations themselves can be deduced. To find the MO's belonging to a given representation one begins with an AO and notes its transformation on application of each operation in the group. Then by multiplying by the character of the representation for each operation and summing, one obtains a linear combination of the MO's which belong to that representation.

In proving that a set of AO's can be used to deduce the representations spanned by the MO's formed from them, we found it convenient to introduce matrix notation; and we saw that the rule for finding the element z_{ij} in a matrix $[Z] = [X][Y]$ is to multiply each element x_{ik} in the ith row of $[X]$ by the element y_{kj} in the jth row of $[Y]$ and to sum.

We next turned our attention to the effect of the requirement that electronic wavefunctions have an eigenvalue of -1 on a permutation of two electrons. In order that our electronic wavefunctions possess the required antisymmetrization, we found it expedient to write them as Slater determinants. From the properties of determinants, we were able to show that our many-electron wavefunctions would obey the exclusion principle and that we could form combinations of the orbitals within a determinant. We saw that the latter process can be used to attempt to find orbitals that, while not being eigenfunctions of certain operators, do give a localized description of bonding.

Finally, we discussed the Born-Oppenheimer approximation, which permits us to deal solely with the electronic wavefunction by allowing us to separate it from the wavefunction for the nuclei. However, we saw that the symmetry of the nuclear wavefunctions could be used in the description of vibronically induced orbital mixing.

FURTHER READING

F. A. COTTON, *Chemical Applications of Group Theory*, Interscience, New York, 1963. The author states: "I have attempted to make this the kind of book which one can read in bed without a pencil. . .." A second edition has recently appeared which is somewhat more sophisticated and hence somewhat less suitable for bedtime reading than the first.

ROBIN M. HOCHSTRASSER, *Molecular Aspects of Symmetry*, W. A. Benjamin, New York, 1966. This book is more theoretically oriented and contains fewer examples of applications to problems of obvious relevance to an organic chemist.

R. G. PEARSON, *J. Amer. Chem. Soc.* **91**, 4947 (1969), applies the second-order Jahn-Teller effect to the prediction of molecular geometry. Other applications are given in a review in *Accts. Chem. Res.* **4**, 152 (1971).

7

Explicit Inclusion of Electron Repulsion in the Hamiltonian— Pariser-Parr-Pople Theory

We have seen the desirability of explicitly including electrostatic terms, particularly that for electron–electron repulsion, in the Hamiltonian; and having learned in the last chapter how to write many-electron wavefunctions correctly, we are now ready to take this step. In so doing we will utilize some approximations, which can reasonably be made, in order to give us a semiempirical method that is only slightly less convenient to use than Hückel theory. The resulting theory is known as the Pariser-Parr-Pople (PPP) method, named for the theoreticians responsible for its development.

The π Electron Hamiltonian

When the electron repulsion operator e^2/r_{ij} is included in the Hamiltonian, the sigma wavefunction affects the pi and vice versa, since the electrons in the sigma and pi bonds tend to distribute themselves so as to minimize their repulsive interaction. Therefore, we cannot find the two many-electron wavefunctions, Ψ_σ and Ψ_π, independently when the Hamiltonian contains an interaction term, $\mathcal{3C}_{\sigma\pi}$. Nevertheless, suppose we solve $\mathcal{3C}_\sigma \Psi_\sigma = E_\sigma \Psi_\sigma$ to determine an approximate Ψ_σ. We can use this in the equation

$$\mathcal{3C} \Psi_\sigma \Psi_\pi = E \Psi_\sigma \Psi_\pi \tag{1}$$

to determine a Ψ_π, where $\mathcal{3C}$ is the full Hamiltonian $\mathcal{3C} = \mathcal{3C}_\sigma + \mathcal{3C}_\pi + \mathcal{3C}_{\sigma\pi}$. The Ψ_π we find will not, however, be quite the correct pi wavefunction; for it was determined with an approximate Ψ_σ. But it can now be used in (1) to determine a better Ψ_σ and the latter then used to get an even better approximation to Ψ_π, until the Ψ_π used in the calculation of Ψ_σ becomes the same as the Ψ_π calculated from the Ψ_σ. At this point the calculation is said to have reached self-consistency; and in the next chapter we shall see in detail the importance of self-consistent methods in finding correct wavefunctions when the two-electron operator, e^2/r_{ij}, is explicitly included in the Hamiltonian.

Let us assume now that we somehow know the correct Ψ_σ^o, where the superscript signifies that it is the lowest energy sigma wavefunction for a system, and use it and the full Hamiltonian to determine the corresponding Ψ_π^o. We can partition the Hamiltonian into $\mathcal{3C}_\sigma$ and $\mathcal{3C}_\pi'$ by defining $\mathcal{3C}_\pi' = \mathcal{3C}_\pi + \mathcal{3C}_{\sigma\pi}$. Therefore, with no loss of rigor we can still write

$$(\mathcal{3C}_\sigma + \mathcal{3C}_\pi') \Psi_\sigma^o \Psi_\pi^o = (E_\sigma^o + E_\pi^o) \Psi_\sigma^o \Psi_\pi^o \tag{2}$$

so that all the interaction energy is contained in E_π^o. If we now make the assumption that the nature of the sigma-pi interaction is such that were we to recalculate Ψ_σ^o for any one of a number of pi states, Ψ_π^i, of interest to us, it would remain the same as Ψ_σ^o for the pi ground state, (2) can be generalized to any pi state, Ψ_π^i. Subtracting $\mathcal{3C}_\sigma \Psi_\sigma^o \Psi_\pi^i = E_\sigma^o \Psi_\sigma^o \Psi_\pi^i$ from the generalized (2) yields

$$\mathcal{3C}' \Psi_\sigma^o \Psi_\pi^i = (\mathcal{3C}_\pi + \mathcal{3C}_{\sigma\pi}) \Psi_\sigma^o \Psi_\pi^i = E_\pi^i \Psi_\sigma^o \Psi_\pi^i \tag{3}$$

Multiplication by Ψ_σ^o and integration over the sigma wavefunction transforms (3) into

$$(\mathcal{3C}_\pi + V_{\sigma\pi}) \Psi_\pi^i = E_\pi^i \Psi_\pi^i \tag{4}$$

where $V_{\sigma\pi} = \int \Psi_\sigma^o \mathcal{H}_{\sigma\pi} \Psi_\sigma^o \, d\tau_\sigma$ is the potential, due to the sigma electrons, that acts on the pi electrons. Therefore, use of what is known as the "π electron approximation"—that the sigma wavefunction remains the same for any state of the pi system—simplifies the problem of finding the wavefunctions for a planar unsaturated molecule to that of solving, for just the π system, a Schrödinger equation in which the σ electrons are treated as making a constant contribution to the potential in which the π electrons move. We anticipate that the π electron approximation can be expected to work best for molecules where the π electron distribution remains pretty much the same for all states of interest, so that the σ wavefunction will be the same for all these π states.

Having made the π electron approximation, the Schrödinger equation for the π system can be written

$$\left(-\frac{\hbar^2}{2m} \sum_i \nabla_i^2 + V\right)\Psi = E\Psi \tag{5}$$

where we have dropped the π subscripts, since we will deal exclusively with π electrons in this and the following two chapters. The first term in the Hamiltonian is a differential operator which gives the kinetic energy of the electrons. The potential term in the π electron Hamiltonian is

$$V = \sum_{A>B,B} \frac{Z_A' Z_B' e^2}{r_{AB}} + \sum_{i>j,j} \frac{e^2}{r_{ij}} - \sum_{A,i} \frac{Z_A' e^2}{r_{Ai}} \tag{6}$$

where Z_A' is the number of pi electrons that atom A contributes, since we assume that the effect of the sigma valence electrons is to cancel out all but this number of the nuclear positive charges. The terms in the potential are respectively the nuclear-nuclear repulsion, the electron–electron repulsion, and the nuclear-electron attraction. The Hamiltonian is not, therefore, as we have tacitly been assuming, a one-electron operator. The presence of the term e^2/r_{ij} means that, in order to find the wavefunction for one π electron, we must know the wavefunctions of all the other π electrons. Therefore, we cannot use (5) to calculate a one-electron wavefunction (i.e., an MO) unless we already know the wavefunctions for all the filled MO's in the molecule. This means that we will have to use self-consistent methods to obtain the MO's, when they are not determined by symmetry; and we will defer discussion of this topic until the next chapter.

Here we will confine ourselves to a discussion of the effects of the inclusion of electron repulsion in the Hamiltonian on the treatment of molecules where the MO's are determined by symmetry.

Ground State of Ethylene

In ethylene, as we have seen, the MO's are determined by symmetry. We can compute the energy of the ground state $\Psi = |\psi_1^\alpha(1)\psi_1^\beta(2)\rangle$, where both electrons are in the bonding MO, ψ_1, from

$$E = \int \Psi^* \mathcal{H} \Psi \, d\tau \equiv \langle \psi_1^\alpha(1)\psi_1^\beta(2) \, | \, \mathcal{H} \, | \, \psi_1^\alpha(1)\psi_1^\beta(2)\rangle \qquad (7)$$

where we introduce bra-ket notation as a shorthand to denote an integral over the space and spin coordinates of all the electrons. In evaluating (7), it is convenient to divide the terms in the Hamiltonian into three types: those which operate on zero, on one, and on two electrons at a time. In the first category is the term that represents the nuclear repulsion, and it just gives an additive constant to the energy of any state.

$$\left\langle \psi_1^\alpha \psi_1^\beta \, \middle| \, \sum_{A>B,B} \frac{Z_A' Z_B' e^2}{r_{AB}} \, \middle| \, \psi_1^\alpha \psi_1^\beta \right\rangle = \frac{e^2}{r_{AB}} \langle \psi_1^\alpha \psi_1^\beta \, | \, \psi_1^\alpha \psi_1^\beta \rangle = \frac{e^2}{r_{AB}} \qquad (8)$$

The last equality follows from the fact that the Slater determinants, represented by the bra and ket, are normalized.

To the second class belong the kinetic energy and nuclear-electron attraction operators, which we will together denote $\hbar$. In order to see what terms these one-electron operators contribute to the energy, let us expand the Slater determinants represented by the bra and ket in (7). In general, we shall want scrupulously to avoid actually expanding such Slater determinants, since the number of terms in the expansion goes as $n!$. Only for a two-electron problem is the expansion convenient, and we do it here only to demonstrate convincingly that the answers found by expansion can be much more readily obtained directly from the diagonals of the Slater determinants, which we use in bras and kets to represent the determinants themselves. Expanding the integral over the one-electron operator $\hbar = \hbar(1) + \hbar(2)$, where $\hbar(1)$ operates only on electron 1 and $\hbar(2)$ only on 2, yields

$$\langle \psi_1^\alpha \psi_1^\beta | \hbar | \psi_1^\alpha \psi_1^\beta \rangle \equiv \langle \psi_1^\alpha(1)\psi_1^\beta(2) | \hbar(1) + \hbar(2) | \psi_1^\alpha(1)\psi_1^\beta(2) \rangle$$

$$= \int \frac{1}{\sqrt{2}} [\psi_1^\alpha(1)\psi_1^\beta(2) - \psi_1^\beta(1)\psi_1^\alpha(2)]$$

$$\times [\hbar(1) + \hbar(2)] \frac{1}{\sqrt{2}} [\psi_1^\alpha(1)\psi_1^\beta(2)$$

$$- \psi_1^\beta(1)\psi_1^\alpha(2)] \, d\tau \qquad (9)$$

$$= \frac{1}{2} \int \{ [\psi_1^\alpha(1)\psi_1^\beta(2)][\hbar(1) + \hbar(2)][\psi_1^\alpha(1)\psi_1^\beta(2)]$$

$$+ [\psi_1^\beta(1)\psi_1^\alpha(2)][\hbar(1) + \hbar(2)][\psi_1^\beta(1)\psi_1^\alpha(2)]$$

$$- [\psi_1^\alpha(1)\psi_1^\beta(2)][\hbar(1) + \hbar(2)][\psi_1^\beta(1)\psi_1^\alpha(2)]$$

$$- [\psi_1^\beta(1)\psi_1^\alpha(2)][\hbar(1) + \hbar(2)][\psi_1^\alpha(1)\psi_1^\beta(2)] \} \, d\tau$$

The integrals in (9) are over both the space and spin coordinates of the electrons. Since the operators have no effect on the spin functions, the integrals over the spins are easiest to carry out; and we shall do them first. The rules for these integrals are

$$\int \alpha(1)\alpha(1) = \int \beta(1)\beta(1) = 1, \qquad \int \alpha(1)\beta(1) = 0 \qquad (10)$$

We note immediately that the terms in (9) that are the product of the diagonal elements of one determinant and the off-diagonal elements of the other, and which, consequently, occur with a minus sign in (9), are all zero because they involve integrals of the second type. Thus, (9) reduces to the terms that are products of diagonal elements, because they have spin functions that integrate to one. Let us, however, postpone actually carrying out the integration over those spin functions and, instead, simply expand further the nonzero terms. At the same time we will make use of the fact that $\hbar(1)$ only operates on electron 1 and $\hbar(2)$ on 2 to divide the integrals over the coordinates of both electrons into integrals over those of each electron separately. The result is

$$\langle \psi_1^\alpha \psi_1^\beta | \hbar | \psi_1^\alpha \psi_1^\beta \rangle = \frac{1}{2} \Big\{ \Big[\int \psi_1^\alpha(1)\hbar(1)\psi_1^\alpha(1) \, d\tau\,(1) \Big]\Big[\int \psi_1^\beta(2)\psi_1^\beta(2) \, d\tau\,(2) \Big]$$

$$+ \Big[\int \psi_1^\alpha(1)\psi_1^\alpha(1) \, d\tau\,(1) \Big]\Big[\int \psi_1^\beta(2)\hbar(2)\psi_1^\beta(2) \, d\tau\,(2) \Big]$$

$$+ \Big[\int \psi_1^\beta(1)\hbar(1)\psi_1^\beta(1) \, d\tau\,(1) \Big]\Big[\int \psi_1^\alpha(2)\psi_1^\alpha(2) \, d\tau\,(2) \Big]$$

$$+ \Big[\int \psi_1^\beta(1)\psi_1^\beta(1) \, d\tau\,(1) \Big]\Big[\int \psi_1^\alpha(2)\hbar(2)\psi_1^\alpha(2) \, d\tau\,(2) \Big] \Big\}$$

$$(11)$$

The first and last terms in (11) are identical, except for the way that the electrons are labelled, as are the two middle terms. Since electrons are indistinguishable, which we call 1 and which we call 2 is physically meaningless. Thus, (11) can equivalently be written

$$\langle \psi_1^\alpha \psi_1^\beta | \hbar | \psi_1^\alpha \psi_1^\beta \rangle = \left[\int \psi_1^\alpha(1) \hbar(1) \psi_1^\alpha(1) \, d\tau \,(1) \right] \left[\int \psi_1^\beta(2) \psi_1^\beta(2) \, d\tau \,(2) \right]$$
$$+ \left[\int \psi_1^\alpha(1) \psi_1^\alpha(1) \, d\tau \,(1) \right] \left[\int \psi_1^\beta(2) \hbar(2) \psi_1^\beta(2) \, d\tau \,(2) \right]$$

$$(12)$$

Using the definition of $\hbar$ in the left hand side of (12) and factoring the right hand side we finally obtain

$$\langle \psi_1^\alpha(1) \psi_1^\beta(2) | \hbar(1) + \hbar(2) | \psi_1^\alpha(1) \psi_1^\beta(2) \rangle$$
$$= \int \psi_1^\alpha(1) \psi_1^\beta(2) [\hbar(1) + \hbar(2)] \psi_1^\alpha(1) \psi_1^\beta(2) \, d\tau$$

$$(13)$$

Equation (13) shows that the integral over the Slater determinants gives the same result as just integrating over the products of their diagonal elements, by which the Slater determinants are represented in bra-ket notation. Why is this the case?

When we actually expanded the Slater determinants in (9) we obtained four terms. The first came from the diagonals of both determinants. The second came from the off-diagonal elements in both determinants, so that the part coming from the bra and the part from the ket had the electron labels permuted identically. This meant that two orbitals (products of space and spin parts) which were integrated together in the first term in (9) were also integrated together in the second. This resulted in the first and fourth terms in (11) being identical, as were the second and third. The fact that we had two sets of two identical terms inside the braces in (11) led to the cancellation of the factor of $\frac{1}{2}$, outside the braces, which arose from the normalization of the Slater determinants. This is what happens whenever one does integrals over two Slater determinants—one obtains $n!$ terms which are identical (except for the physically irrelevant labelling of the electrons) and which arise from the $n!$ identical permutations of the diagonal elements of the Slater determinants in the bra and ket. Since these $n!$ identical terms are multiplied by a factor of $1/\sqrt{n!} \times 1\sqrt{n!} = 1/n!$, from the normalization of the Slater determinants, only one term is obtained from them. As Equation (13) shows, this term is just the one that appears in the bra-ket.

This does not mean, however, that the term that appears in the bra-ket will always be the only nonzero one in an integral over two Slater determinants. As we shall see in the next section, there can be other non-zero terms, which are usually encountered in integrating Slater determinants over the two-electron, Coulomb repulsion operator. These other terms arise from nonidentical permutations of the diagonal elements of the two Slater determinants. In Equation (9) such terms appeared with a minus sign and came from the diagonal elements of one determinant and the off-diagonal elements of the other. They integrated to zero because they involved orthogonal spin functions. Suppose, however, that both one-electron wavefunctions in our bra and in our ket had the same spin; would these terms then be nonzero? The answer is that they would still give zero, for the following reason. If the two one-electron wavefunctions have the same spin, the exclusion principle tells us that they must have different spatial parts, or the Slater determinant representing the many-electron wavefunction will vanish. These different spatial one-electron wavefunctions are just two different MO's, which are, of course, orthogonal. Because they are orthogonal, on carrying out the integrals over the spatial parts of the wavefunction, one again finds that the terms arising from different permutations of the diagonal elements vanish. Thus, *integrals of one-electron operators over Slater determinants* are trivial to evaluate, since they *are just equal to the integral of the term that appears in the bra-ket.*

Problem 7.1. Instead of the integral $\langle \psi_1^\alpha \psi_1^\beta | \hbar | \psi_1^\alpha \psi_1^\beta \rangle$ consider $\langle \psi_1^\alpha \psi_2^\alpha | \hbar | \psi_1^\alpha \psi_2^\alpha \rangle$. Without expanding the Slater determinants, write down what you expect this second bra-ket to equal. Check your answer by expanding the Slater determinants. Show that the terms arising from different permutations of the diagonal elements are, indeed, zero, despite the fact that the spin functions in these terms integrate to unity.

On completing the integrals over space and spin in (12) we obtain

$$\langle \psi_1^\alpha \psi_1^\beta | \hbar | \psi_1^\alpha \psi_1^\beta \rangle = \int \psi_1(1) \hbar(1) \psi_1(1) \, d\tau \, (1) + \int \psi_1(2) \hbar(2) \psi_1(2) \, d\tau \, (2)$$
$$= 2\langle \psi_1 | \hbar | \psi_1 \rangle \tag{14}$$

where we have taken advantage of the identity

$$\int \psi_1(1) \hbar(1) \psi_1(1) \, d\tau \, (1) = \int \psi_1(2) \hbar(2) \psi_1(2) \, d\tau \, (2) \equiv \langle \psi_1 | \hbar | \psi_1 \rangle \tag{15}$$

Equation (14) shows that *the one-electron energy of a ket is equal to the sum of the energies of the one-electron wavefunctions which comprise it.*

Let us now turn (without neglecting overlap for the moment) to the actual evaluation of the integral

$$\langle \psi_1 | \hbar | \psi_1 \rangle = \frac{1}{2(1+S)} \left\langle (\phi_A + \phi_B) \left| -\frac{\hbar^2}{2m} \nabla_i^2 - \frac{e^2}{r_{Ai}} - \frac{e^2}{r_{Bi}} \right| (\phi_A + \phi_B) \right\rangle \tag{16}$$

Since the two π AO's in ethylene are identical, this can be simplified to

$$\langle \psi_1 | \hbar | \psi_1 \rangle = \frac{1}{1+S} \left[\left\langle \phi_A \left| -\frac{\hbar^2}{2m} \nabla_i^2 - \frac{e^2}{r_{Ai}} \right| \phi_A \right\rangle + \left\langle \phi_A \left| -\frac{e^2}{r_{Bi}} \right| \phi_A \right\rangle \right. $$
$$\left. + \left\langle \phi_A \left| -\frac{\hbar^2}{2m} \nabla_i^2 - \frac{e^2}{r_{Ai}} - \frac{e^2}{r_{Bi}} \right| \phi_B \right\rangle \right] \tag{17}$$

The first of these terms represents the total energy of an electron occupying a $p\text{-}\pi$ orbital on an isolated atom; hence, this term can be set equal to $-I_p$, the negative of the valence state ionization potential for an electron in such an orbital. The second term is the attraction of the effective nuclear charge of atom B for the electron on A. If we assume that without its $p\text{-}\pi$ electron, the distribution of positive charge on B is essentially the positive hole left when the electron is removed, this term becomes $-\gamma_{AB}$, the attraction of a positron in the p orbital on B for an electron in the p orbital on A. The final term represents the energy of the electron density in the overlap region, which we will again denote by β. Thus,

$$\langle \psi_1 | \hbar | \psi_1 \rangle = \frac{-I - \gamma_{AB} + \beta}{1+S} = -I - \gamma_{AB} + \beta' \tag{18}$$

where we again have carried out a procedure similar to that of Chapter 1 for ridding ourselves of S and will henceforth drop the prime on β.

Although we have noted that, as in Hückel theory, the one-electron energies in ethylene add, there is only one term arising from the two-electron operator

$$\left\langle \psi_1^\alpha \psi_1^\beta \left| \frac{e^2}{r_{12}} \right| \psi_1^\alpha \psi_1^\beta \right\rangle = \int \psi_1^\alpha(1) \psi_1^\alpha(1) \frac{e^2}{r_{12}} \psi_1^\beta(2) \psi_1^\beta(2) \, d\tau \tag{19}$$

since the term arising from the diagonal of one determinant and the off-diagonal of the other is again zero, due to an integral over orthogonal spin functions. We now define the symbol $(\psi_i\psi_j|\psi_k\psi_l)$ as the repulsion integral between the two charge distributions given by $e\psi_i(1)\psi_j(1)$ and $e\psi_k(2)\psi_l(2)$. After integration over the spins, (19) in this convenient notation becomes

$$\left\langle \psi_1^\alpha\psi_1^\beta \middle| \frac{e^2}{r_{12}} \middle| \psi_1^\alpha\psi_1^\beta \right\rangle = (\psi_1\psi_1|\psi_1\psi_1) \equiv J_{11} \qquad (20)$$

The Coulomb integral, J_{11}, is just equal to the electrostatic repulsion between two identical charge distributions, $e\psi_1\psi_1$, given by

$$e\psi_1\psi_1 = \frac{e}{2+2S}[\phi_A^2 + \phi_B^2 + 2\phi_A\phi_B] \qquad (21)$$

When evaluated this is, again using the fact that the AO's are identical,

$$J_{11} = \frac{2}{(2+2S)^2}[(\phi_A^2|\phi_A^2) + (\phi_A^2|\phi_B^2) + 2(\phi_A\phi_B|\phi_A\phi_B) + 4(\phi_A\phi_B|\phi_A^2)] \qquad (22)$$

The four terms in (22) represent, respectively, the electrostatic repulsion between two electrons: (1) on the same atom, (2) on different atoms, (3) both in the overlap region, and (4) one in the overlap region and one on an atom. The number of such integrals becomes very large very rapidly in a molecule of any size; in fact, it goes as the number of AO's to the fourth power. Therefore, if the inclusion of electron repulsion is not to cause tremendous algebraic complications in our calculations, we must somehow get rid of the terms that involve overlap distributions and that are largely responsible for the n^4 problem.

In Chapter 1 we used the Mulliken approximation to an overlap distribution to suggest a means of eliminating the overlap integral in the denominator of our one-electron energy expressions; let us now apply it to the rewriting of the overlap distributions in (22). Substituting

$$\phi_A\phi_B = \tfrac{1}{2} S [\phi_A^2 + \phi_B^2] \qquad (23)$$

we obtain

$$
\begin{aligned}
J_{11} &= \frac{2}{(2 + 2S)^2}\Big[(\phi_A^2 \,|\, \phi_A^2) + (\phi_A^2 \,|\, \phi_B^2) \\
&\quad + \frac{S^2}{2}(\phi_A^2 + \phi_B^2 \,|\, \phi_A^2 + \phi_B^2) + 2S(\phi_A^2 + \phi_B^2 \,|\, \phi_A^2) \Big] \\
&= \frac{1}{2(1 + S)^2}[(\phi_A^2 \,|\, \phi_A^2)(1 + 2S + S^2) \\
&\quad + (\phi_A^2 \,|\, \phi_B^2)(1 + 2S + S^2)] \\
&= \frac{1}{2}[(\phi_A^2 \,|\, \phi_A^2) + (\phi_A^2 \,|\, \phi_B^2)] = \frac{1}{2}[\gamma_{AA} + \gamma_{AB}]
\end{aligned}
\tag{24}
$$

Thus, use of the Mulliken approximation allows the *formal* neglect of not only the integral of the overlap, but also of the fact that the orbitals overlap at all; since the above value of J_{11} is what we would calculate from (22) if we assumed *zero differential overlap* (i.e., $\phi_A\phi_B = 0$ and hence, $S = 0$ also).

Problem 7.2.   Without neglecting overlap, calculate J_{22}, the repulsion between two electrons in the antibonding ethylene MO. Show that with the Mulliken approximation it is the same as J_{11}. Do the same for J_{12}. Show that these are the same values one would calculate assuming zero differential overlap.

Let us examine why use of the Mulliken approximation allows the formal neglect of overlap between AO's on different atoms. Returning to the case of ethylene, if we ask for the relative probabilities of finding an electron on A, on B, and in the overlap region, these are given by

$$
\int \psi_1^2 \, d\tau = \int (\phi_A + \phi_B)^2 \, d\tau = \int \phi_A^2 \, d\tau + \int \phi_B^2 \, d\tau + 2 \int \phi_A\phi_B \, d\tau
\tag{25}
$$

as $1 : 1 : 2S$. Since the total probability must add up to unity, to obtain the absolute probability of finding an electron in one of these three regions, we must divide the relative probabilities by their sum, $2 + 2S$. This is, of course, the origin of our normalization of our wavefunctions. Now suppose that we wish to partition the overlap region back into contributions from the two atoms, which the Mulliken approximation allows us to do. The probabilities of finding the electrons in the various regions of interest are unchanged by this process, but we can now determine from

$$
\int \psi_1^2 \, d\tau = \frac{1}{2 + 2S} \left[\int \phi_A^2 \, d\tau + \int \phi_B^2 \, d\tau + S\left(\int \phi_A^2 \, d\tau + \int \phi_B^2 \, d\tau \right) \right]
$$

$$
= \frac{1}{2(1 + S)} \left[(1 + S) \int \phi_A^2 \, d\tau + (1 + S) \int \phi_B^2 \, d\tau \right] \qquad (26)
$$

$$
= \frac{1}{2} \left[\int \phi_A^2 \, d\tau + \int \phi_B^2 \, d\tau \right]
$$

that the total probability of finding the electron somewhere within ϕ_A is $\frac{1}{2}$. Therefore, by applying the Mulliken approximation we can formally neglect differential overlap, because we are partitioning the overlap region back into contributions from the constituent atoms. We can thus avoid calculating electron repulsions due to charge distributions in the overlap region by including in the charge density for the atom itself its contribution to the overlap charge density.

Problem 7.3. Use the wavefunction $\psi = 3\phi_1 + \phi_2$ for the bonding MO of the heteronuclear molecule of Problem 2.10 to find the absolute probabilities of finding an electron on atom 1, on atom 2, and in the overlap region. Partition the overlap distribution using the Mulliken approximation. Does this partitioning make sense in terms of the actual contributions of the two atoms to the overlap charge? Show how you would partition the overlap region so that the probability of an electron being in each of the AO's is the same as that which would be calculated by assuming zero differential overlap.

With the approximation that we have made regarding the treatment of the nuclear charge as a positron in a p orbital, the total energy of the ground state of ethylene can be written

$$
E = \frac{e^2}{r_{AB}} + 2\langle \psi_1 | \hbar | \psi_1 \rangle + J_{11}
$$

$$
= \gamma_{AB} - 2I + 2\beta - 2\gamma_{AB} + \frac{1}{2}(\gamma_{AA} + \gamma_{AB}) \qquad (27)
$$

$$
= -2I + 2\beta + \frac{1}{2}(\gamma_{AA} - \gamma_{AB})
$$

This is identical to Equation (3) of Chapter 5, which we obtained by direct analysis of the charge distribution given by the MO wavefunction for two equivalent atoms.

As before we will treat β as a parameter to be derived from experiment, and we now turn to the evaluation of γ_{AA} and γ_{AB}. These can be computed theoretically using Slater orbitals, and the calculated values are $\gamma_{AA} = 16.93$ eV and $\gamma_{AB} = 9.24$ eV. As we saw in Problem 10 of Chapter 1, however, γ_{AA} is equivalent to the energy required to ionize a carbon atom in its valence state and transfer the electron to a second neutral carbon atom at infinity. The valence state ionization potential and electron affinity of carbon can be determined from spectroscopic data. A semiempirical value for γ_{AA} can thus be found as

$$\gamma_{AA} = I - \alpha = 11.54 - 0.46 = 11.08 \text{ eV} \qquad (28)$$

Why is there a difference of nearly 6 eV between the theoretical and semiempirical values for this integral? One reason is suggested by the virial theorem. When two electrons occupy the same orbital, although the potential energy increases, this effect will be somewhat offset by a decrease in the kinetic energy of each electron.[1] The above theoretical evaluation of γ_{AA} does not take into account possible changes in the orbital size. Moreover, the purely theoretical value is derived by assuming that when both electrons occupy the same carbon $2p$ orbital, their repulsion is just that found by superimposing two charge density distributions calculated from a Slater $2p$ orbital. But physically this makes little sense, for we know that two electrons will tend to correlate their motions in order to avoid each other, so that, for example, when one electron is in one lobe of a p orbital, a second electron will tend to occupy the other. Our failure to take account of this correlation of motions in computing γ_{AA} will also make our theoretical value too high. Therefore, if we hope to be able to make quantitatively correct predictions, we are obliged to use the semiempirical value for γ_{AA}.

Numerous schemes have been proposed for effecting similar reductions from the purely theoretical values for two-center integrals of the γ_{AB} type. We might expect that for atoms which are not nearest neighbors,

---

[1] The physical reason for this is that when two electrons simultaneously occupy the same AO, their repulsion is diminished if the orbital expands. Since the electrons then have a bigger "box" in which to move, their kinetic energy also decreases.

correlation effects would be small—and this is the case. For instance, in benzene the theoretically calculated repulsion integrals for orbitals *meta* and *para* to each other are very close to those which may be derived semi-empirically from the spectrum of benzene; only for orbitals *ortho* to each other is a reduction from the theoretical value necessary. For orbitals that are *not* nearest neighbors it is convenient to use Parr's uniformly charged spheres, which give almost the same values for electron repulsion energies as do the much-more-difficult-to-perform integrals over Slater orbitals. In the uniformly charged spheres model, the repulsion between two electrons in p orbitals on different atoms is set equal to the repulsion of half-unit charges in tangent spheres, of diameter D = 1.45 Å, located above and below each nucleus. Since the potential due to a uniformly charged sphere is the same as if all the charge were concentrated at its center, the calculation reduces to one involving point charges of one half electron located at D/2 above and below each atom. Therefore, for non-nearest neighbors, γ_{AB} can be written analytically as

$$\gamma_{AB} = \frac{e^2}{2r_{AB}}\left(1 + \frac{1}{\sqrt{1 + (D/r_{AB})^2}}\right) \tag{29}$$

where $e^2 = 14.4$ eV-Å.

An extension, suggested by Dewar, of the uniformly charged spheres approximation allows for electron correlation effects by assuming that repulsions are entirely between electrons on opposite sides of the nodal plane. An electron repulsion integral between any two atoms is then calculated from

$$\gamma_{AB} = \frac{14.4}{\sqrt{D^2 + r_{AB}^2}}\,\text{eV} \tag{30}$$

D can be chosen as 1.30 Å to fit the semiempirical value of the one-center repulsion integral, $\gamma_{AA} = 11.1$ eV.

Excited States of Ethylene

Let us now turn to the actual determination of a semiempirical value for γ_{AB} in ethylene. In order to find this number we can use the spectrum of ethylene and compare the theoretical expressions for the energy of the $\pi \rightarrow \pi^*$ singlet and triplet states, relative to the ground state, with the spectroscopically derived values. It is possible to write four Slater determinants for a one-electron excitation in ethylene

$$|\psi_1^\alpha \psi_2^\alpha\rangle \qquad |\psi_1^\beta \psi_2^\beta\rangle \qquad |\psi_1^\alpha \psi_2^\beta\rangle \qquad |\psi_1^\beta \psi_2^\alpha\rangle \qquad (31)$$

The first two have values of S_z of $+1$ and -1, respectively; so they must be components of the triplet. The energy of these two configurations is given by

$$E = \gamma_{AB} + \langle \psi_1 | \hbar | \psi_1 \rangle + \langle \psi_2 | \hbar | \psi_2 \rangle + \left\langle \psi_1^\alpha \psi_2^\alpha \left| \frac{e^2}{r_{12}} \right| \psi_1^\alpha \psi_2^\alpha \right\rangle \qquad (32)$$

Expansion of the determinants in the electron repulsion integral yields

$$\left\langle \psi_1^\alpha \psi_2^\alpha \left| \frac{e^2}{r_{12}} \right| \psi_1^\alpha \psi_2^\alpha \right\rangle = \int \psi_1^\alpha(1)\psi_1^\alpha(1) \frac{e^2}{r_{12}} \psi_2^\alpha(2)\psi_2^\alpha(2) \, d\tau$$

$$- \int \psi_1^\alpha(1)\psi_2^\alpha(1) \frac{e^2}{r_{12}} \psi_1^\alpha(2)\psi_2^\alpha(2) \, d\tau \qquad (33)$$

$$= (\psi_1\psi_1 | \psi_2\psi_2) - (\psi_1\psi_2 | \psi_1\psi_2)$$

where the second integral arises because the product of the diagonal term in one determinant and the off-diagonal term in the other gives an integral of the two-electron operator over spin and space wavefunctions which is not zero, as it was for the ethylene ground state; since here the electrons have the same spin. The first of the terms in (33) is just J_{12}; the second is called the exchange integral K_{12} and corrects for the fact that, because the electrons have the same spin, they tend to avoid each other.

Problem 7.4.   Verify (33) by expansion of the Slater determinants. Compute J_{12} and K_{12} in terms of γ_{AA} and γ_{AB} and show, from the net value of the electron repulsion energy in the triplet state, that the correlation of the motions of electrons of the same spin (required by the exclusion principle via the antisymmetrization of the total electronic wavefunction) forces electrons of the same spin always to be in different AO's.

Equation (33) is the first example that we have encountered where nonzero integrals arise from terms other than the one obtained from the diagonals of both Slater determinants. We have already seen that we need not expand Slater determinants fully to calculate integrals when the only nonzero ones arise from the $n!$ identical permutations of the diagonal terms. We will now show that expansion is also unnecessary in order to find which of the $n! - 1$ types of $n!$ identical integrals, arising from different permutations of the diagonal terms in the determinants, is nonzero. The reason that we need not expand the determinants is that we can always rewrite one of them in such a way that an off-diagonal term, which represents one of the permutations of the diagonal, becomes the diagonal one. This just involves switching rows in the determinant, remembering that every time we interchange two rows, the determinant changes sign. Thus, after switching two rows of the ket, (33) can equivalently be written

$$\left\langle \psi_1^\alpha \psi_2^\alpha \left| \frac{e^2}{r_{12}} \right| \psi_1^\alpha \psi_2^\alpha \right\rangle = -\left\langle \psi_1^\alpha \psi_2^\alpha \left| \frac{e^2}{r_{12}} \right| \psi_2^\alpha \psi_1^\alpha \right\rangle \tag{34}$$

Both of the terms that occur on actually expanding the Slater determinants in the bra-ket in (33) are revealed in (34). In any bra-ket we can obtain all the possible integrals, which arise on expansion of the Slater determinants, through rewriting one of the determinants in the $n! - 1$ other equivalent ways possible by interchanging rows. We need not, however, examine all $n!$ possible permutations of rows since many of these will give integrals that are zero. Any integral over a two-electron operator in which three or more electrons have different spatial wavefunctions will be zero, since it will involve an integral of the type $\int \psi_i \psi_j \, d\tau$, which is zero by orthogonality. For a one-electron operator only two spatial wavefunctions need differ in order to ensure a zero. Of course, a difference in spin function for an electron in the bra and the same one in the ket will also cause an integral to be zero.

Problem 7.5.   Obtain the energy of the ground state wavefunction $\Psi = |\psi_1^\alpha \psi_1^\beta \psi_2^\alpha \psi_2^\beta\rangle$ in butadiene in terms of the one-electron integrals $\langle \psi_1 | \hbar | \psi_1 \rangle$ and $\langle \psi_2 | \hbar | \psi_2 \rangle$ and the two-electron integrals J_{11}, J_{22}, J_{12}, and K_{12}.

Returning to the other wavefunctions for the excited configurations of ethylene, there are two Slater determinants in (31) with $S_z = 0$. They both have the same energy.

$$E = \gamma_{AB} + \langle \psi_1 | \hbar | \psi_1 \rangle + \langle \psi_2 | \hbar | \psi_2 \rangle + \left\langle \psi_1^\alpha \psi_2^\beta \left| \frac{e^2}{r_{12}} \right| \psi_1^\alpha \psi_2^\beta \right\rangle \tag{35}$$

However, since they both have the same value of S_z, we have no reason to suppose that they will not mix. To explore this possibility we calculate the value of $\langle \psi_1^\alpha \psi_2^\beta \,|\, \mathcal{K} \,|\, \psi_1^\beta \psi_2^\alpha \rangle$. Because the bra and ket differ in the wavefunctions for two electrons, a nonzero integral can only be obtained for the two-electron operator. The bra-ket thus reduces to

$$\left\langle \psi_1^\alpha \psi_2^\beta \left| \frac{e^2}{r_{12}} \right| \psi_1^\beta \psi_2^\alpha \right\rangle = -\left\langle \psi_1^\alpha \psi_2^\beta \left| \frac{e^2}{r_{12}} \right| \psi_2^\alpha \psi_1^\beta \right\rangle = -(\psi_1 \psi_2 | \psi_2 \psi_1) = -K_{12}$$

(36)

Therefore, the plus combination of the two kets is $-K_{12}$ lower in energy than either alone and has the same energy as the other two components of the triplet. We expect that it is the third component with $S_z = 0$; and in Chapter 9, when we discuss spin operators, we will be able to verify that this is so by operating on its wavefunction with S^2. However, even now we can see that it is a component of the triplet; for we know that if $\psi_1 = \psi_2$, the triplet wavefunction will vanish. It is clear that this is the case for $\Psi = 1/\sqrt{2}(|\psi_1^\alpha \psi_2^\beta\rangle + |\psi_1^\beta \psi_2^\alpha\rangle) = 1/\sqrt{2}(|\psi_1^\alpha \psi_2^\beta\rangle - |\psi_2^\alpha \psi_1^\beta\rangle)$. If $\psi_1 = \psi_2$, the minus combination of the two kets with $S_z = 0$, $\Psi = 1/\sqrt{2}(|\psi_1^\alpha \psi_2^\beta\rangle - |\psi_1^\beta \psi_2^\alpha\rangle) = 1/\sqrt{2}(|\psi_1^\alpha \psi_2^\beta\rangle + |\psi_2^\alpha \psi_1^\beta\rangle)$, does not vanish; and this is the singlet with energy $2K_{12}$ greater than the triplet.

Problem 7.6. It is often said that singlet states have symmetric spatial and antisymmetric spin wavefunctions, while the reverse is true for triplet states. Show that this is the case for the excited ethylene singlet and $S_z = 0$ triplet component by expanding their Slater determinant wavefunctions and factoring the results into spin and space parts.

In the spectrum of ethylene the absorption identified with the N $\rightarrow$ V (ground to excited singlet) transition occurs at 7.6 eV and the N $\rightarrow$ T at 4.6 eV. Setting the theoretical energy difference between these two states equal to the experimental figure,

$$2K_{12} = \gamma_{AA} - \gamma_{AB} = 3.0\,\text{eV}$$

(37)

Using $\gamma_{AA} = 11.1$ eV in (37) gives $\gamma_{AB} = 8.1$ eV—fairly close to the value of 7.7 eV computed from (30) and more than an eV less than the purely theoretical value calculated from Slater orbitals. The mean value of the singlet and triplet state energies, relative to that of the ground state, is $-2\beta = 6.1$ eV, which gives a spectroscopic β of about -3 eV.

There is one more configuration we have yet to consider, the doubly excited one, $|\psi_2^\alpha \psi_2^\beta\rangle$. Now although this electronic configuration is too high in energy to be important spectroscopically, it does have proper symmetry for mixing with the ground state. This mixing turns out to be important in improving the ground state wavefunction for ethylene. Ethylene belongs to the group D_{2h}. If we take x along the molecular axis and z perpendicular to the plane that contains the atoms, the bonding MO belongs to the b_{1u} representation and the antibonding to b_{2g}. We can now determine the representation to which the many-electron wavefunction for each state belongs by determining which representation(s) is spanned by the direct product of the representations to which the orbitals in each belong. In ethylene we get $b_{1u} \times b_{1u} = b_{2g} \times b_{2g} = A_{1g}$ for both $\Psi_1 = |\psi_1^\alpha \psi_1^\beta\rangle$ and $\Psi_2 = |\psi_2^\alpha \psi_2^\beta\rangle$, while all the singly excited states belong to $b_{1u} \times b_{2g} = B_{3u}$.[2]

Since the lowest and the doubly excited configuration in ethylene have the same symmetry and eigenvalues for the electron spin operators, we may anticipate their mixing and compute their interaction. Here again, any integral over a one-electron operator will vanish because the wavefunctions differ in more than one orthogonal MO. However, there is a nonzero integral of the two-electron operator between them:

$$\left\langle \Psi_1 \left| \frac{e^2}{r_{12}} \right| \Psi_2 \right\rangle = \left\langle \psi_1^\alpha \psi_1^\beta \left| \frac{e^2}{r_{12}} \right| \psi_2^\alpha \psi_2^\beta \right\rangle = (\psi_1 \psi_2 | \psi_1 \psi_2) = K_{12} \qquad (38)$$

We can use matrix notation to conveniently display this interaction and to formulate the general problem of finding a better wavefunction for the ethylene ground state. In Chapter 1 we saw that looking for a linear combination of two functions which would minimize the energy led to two equations that we now write in matrix notation as

$$\begin{bmatrix} \langle \Psi_1 | \mathcal{K} | \Psi_1 \rangle & \langle \Psi_1 | \mathcal{K} | \Psi_2 \rangle \\ \langle \Psi_2 | \mathcal{K} | \Psi_1 \rangle & \langle \Psi_2 | \mathcal{K} | \Psi_2 \rangle \end{bmatrix} \begin{bmatrix} c_1 \\ c_2 \end{bmatrix} = \begin{bmatrix} E & 0 \\ 0 & E \end{bmatrix} \begin{bmatrix} c_1 \\ c_2 \end{bmatrix} \qquad (39)$$

[2] The representation to which an MO belongs is usually designated by a small letter. Capitals are reserved for many-electron wavefunctions. One must be aware, in reading the literature, of the choice of axes that an author has made. For instance, the two other possible choices for orientation of the axes would have given B_{2u} and B_{1u} as the representation to which the singly excited states belong.

or equivalently

$$\begin{bmatrix} \langle \Psi_1 | \mathcal{3C} | \Psi_1 \rangle - E & \langle \Psi_1 | \mathcal{3C} | \Psi_2 \rangle \\ \langle \Psi_2 | \mathcal{3C} | \Psi_1 \rangle & \langle \Psi_2 | \mathcal{3C} | \Psi_2 \rangle - E \end{bmatrix} \begin{bmatrix} c_1 \\ c_2 \end{bmatrix} = 0 \qquad (40)$$

so that the integral $\langle \Psi_i | \mathcal{3C} | \Psi_j \rangle$ is really a matrix element of the *Hamiltonian Matrix*, $[H]$. In general, we can define an operator by a series of integrals over the basis functions we are using to construct its eigenfunctions; and, instead of formulating the eigenvalue problem as a differential equation, we can write it as a matrix equation, as in (39). Not surprisingly, in the mathematics of matrices equations like (39), in which we try to find a matrix $[c]$ which diagonalizes $[H]$, are also called eigenvalue problems.

For such problems to have solutions other than the trivial $[c] = 0$, the determinant of the matrix $[H - E]$ must vanish. Thus, the eigenvalues of the Hamiltonian matrix in (39) can be found from the *secular equation*

$$|H - E| = 0 = \begin{vmatrix} H_{11} - E & H_{12} \\ H_{21} & H_{22} - E \end{vmatrix}$$

$$= (H_{11} - E)(H_{22} - E) - H_{12} \cdot H_{21} \qquad (41)$$

which is identical, of course, to the equation we obtain by actually trying to solve two homogeneous equations like (9) and (10) of Chapter 1 for c_1 and c_2. If we take our energy zero at the energy of Ψ_1 so $H_{11} \equiv 0$, then $H_{22} = \langle \Psi_2 | \mathcal{3C} | \Psi_2 \rangle - \langle \Psi_1 | \mathcal{3C} | \Psi_1 \rangle = -4\beta = 12.2$ eV. We have already found that $H_{12} = H_{21} = K_{12} = 1.5$ eV. Substitution in (41) gives $E = -0.2$ eV and $E = 12.4$ eV. The ratio of c_1 to c_2 can be calculated for each of the resulting eigenfunctions from the ratio of the signed minors A_{11} and $-A_{12}$ for each eigenvalue, E. This gives

$$\frac{c_2}{c_1} = -\frac{H_{21}}{(H_{22} - E)} = -0.12$$

for $E = -0.2$ eV.

Why does mixing with the doubly excited configuration improve the ground state wavefunction? We can see by expanding the MO's in the new wavefunction in terms of the AO's which constitute them. We obtain

$$\Psi_1 - 0.12\Psi_2 = \frac{1}{2\sqrt{2}}\{[\phi_A + \phi_B]^\alpha(1)[\phi_A + \phi_B]^\beta(2) - \cdots$$

$$- 0.12[\phi_A - \phi_B]^\alpha(1)[\phi_A - \phi_B]^\beta(2) - \cdots\}$$

$$= \frac{1}{2\sqrt{2}}\{\phi_A^\alpha(1)\phi_A^\beta(2) + \phi_B^\alpha(1)\phi_B^\beta(2) + \phi_A^\alpha(1)\phi_B^\beta(2)$$

$$+ \phi_B^\alpha(1)\phi_A^\beta(2) - \cdots - 0.12[\phi_A^\alpha(1)\phi_A^\beta(2) + \phi_B^\alpha(1)\phi_B^\beta(2) \tag{42}$$

$$- \phi_A^\alpha(1)\phi_B^\beta(2) - \phi_B^\alpha(1)\phi_A^\beta(2) - \cdots]\}$$

$$= \frac{1}{2\sqrt{2}}\{0.88[\phi_A^\alpha(1)\phi_A^\beta(2) + \phi_B^\alpha(1)\phi_B^\beta(2) - \cdots]$$

$$+ 1.12[\phi_A^\alpha(1)\phi_B^\beta(2) + \phi_B^\alpha(1)\phi_A^\beta(2) - \cdots]\}$$

where, for conciseness, we have omitted all the additional terms necessary for antisymmetrization. From the expanded wavefunction we can see that the effect of allowing the doubly excited configuration to interact with the ground state is to reduce the amount of ionic character in the latter from 50% to $(0.88)^2/[(0.88)^2 + (1.12)^2] = 38\%$. This decrease in ionic character improves the MO wavefunction by lowering the high electron repulsion energy concomitant with a large probability of finding two electrons simultaneously on the same atom. In the process, however, the amount of bonding is also lessened so that the net energy decrease is only 0.2 eV. Such mixing of configurations, usually referred to as *configuration interaction*, is expected to be *especially important* in removing ionic character from MO wavefunctions *when the amount of bonding is small*. When two configurations like Ψ_1 and Ψ_2 have the same energy, equal mixing between them occurs; and one of the resulting wavefunctions has all the ionic terms removed from it.

Problem 7.7. Suppose that in a thermal reaction ethylene has rotated 90° about the C—C bond so that the p orbitals are orthogonal. Starting with the wavefunctions for planar ethylene, show that the ground state is purely covalent. Show that the other state of the same symmetry is purely ionic. Are the remaining singly excited triplet and singlet ionic or covalent in nature? Does the ionic character of these latter two states depend on the geometry of the molecule?

Butadiene

Since the MO's of butadiene are not completely determined by symmetry, we cannot actually obtain the best MO's for the molecule, unless we do a self-consistent calculation of the type that we will carry out for butadiene in the next chapter. What we wish to show in butadiene, however, is merely that with electron repulsion included in the Hamiltonian, the electrostatic interaction that produces the difference in excitation energy between the *cisoid* and *transoid* isomers arises naturally. Therefore, let us take the simplest possible set of orthogonal MO's for the problem, namely

$$\psi_{1,3} = \tfrac{1}{2}[(\phi_1 + \phi_4) \pm (\phi_2 + \phi_3)] \quad \psi_{2,4} = \tfrac{1}{2}[(\phi_1 - \phi_4) \pm (\phi_2 - \phi_3)]$$

$$(43)$$

Cisoid butadiene belongs to C_{2v}. The orbitals ψ_1 and ψ_3 form a basis for the b_2 representation while ψ_2 and ψ_4 belong to a_2. Now a transition is allowed if $\mathbf{\mu}_{ij} = e \int \psi_i \mathbf{r} \psi_j \, d\tau \neq 0$. If this is to be the case, we know that the integrand must belong to the totally symmetric representation. Since only the direct product of a representation with itself can give the totally symmetric representation, a vector along one of the coordinate axes must belong to a representation spanned by the product $\psi_i \psi_j$, if the transition is to be electric-dipole-allowed. Of the three lowest possible transitions in butadiene, $\psi_2 \longrightarrow \psi_4$ and $\psi_1 \longrightarrow \psi_3$ belong to A_1, as does $\mathbf{z}$; and $\psi_2 \longrightarrow \psi_3$ belongs to B_1, as does $\mathbf{x}$. Thus, the former two transitions are polarized along the $\mathbf{z}$ axis; and $\psi_2 \longrightarrow \psi_3$ is polarized along $\mathbf{x}$, as we found before from direct analysis of the net transition dipoles in Problem 5.2.

Problem 7.8.   To what group does *transoid* butadiene belong? Are all the same transitions again allowed?

We now proceed to compute the energy of the ground state and of the excited singlet arising from a $\psi_2 \longrightarrow \psi_3$ transition. Since we are really interested only in how the energy of this transition depends on geometry, we can ignore all terms which are independent of geometry (e.g., those involving β, assuming $\beta_{14} = 0$) and those which do not change on excitation (e.g., the nuclear repulsion and the nuclear-electron attraction, since from the pairing theorem the average charge distribution remains the same in this excited state). Thus, we need only compute the terms arising from the two-electron operator. For the ground state, picking out only those terms from the determinants that give rise to nonzero integrals and carrying out the integration over the orbitals in each that are not operated upon, we obtain

$$\left\langle \psi_1^\alpha \psi_1^\beta \psi_2^\alpha \psi_2^\beta \left| \frac{e^2}{r_{12}} \right| \psi_1^\alpha \psi_1^\beta \psi_2^\alpha \psi_2^\beta \right\rangle$$

$$= \left\langle \psi_1^\alpha \psi_1^\beta \left| \frac{e^2}{r_{12}} \right| \psi_1^\alpha \psi_1^\beta \right\rangle + \left\langle \psi_2^\alpha \psi_2^\beta \left| \frac{e^2}{r_{12}} \right| \psi_2^\alpha \psi_2^\beta \right\rangle + \left\langle \psi_1^\alpha \psi_2^\alpha \left| \frac{e^2}{r_{12}} \right| \psi_1^\alpha \psi_2^\alpha \right\rangle$$

$$+ \left\langle \psi_1^\beta \psi_2^\beta \left| \frac{e^2}{r_{12}} \right| \psi_1^\beta \psi_2^\beta \right\rangle + \left\langle \psi_1^\alpha \psi_2^\beta \left| \frac{e^2}{r_{12}} \right| \psi_1^\alpha \psi_2^\beta \right\rangle + \left\langle \psi_1^\beta \psi_2^\alpha \left| \frac{e^2}{r_{12}} \right| \psi_1^\beta \psi_2^\alpha \right\rangle$$

$$= (\psi_1 \psi_1 | \psi_1 \psi_1) + (\psi_2 \psi_2 | \psi_2 \psi_2) + 4(\psi_1 \psi_1 | \psi_2 \psi_2) - 2(\psi_1 \psi_2 | \psi_1 \psi_2)$$

$$= J_{11} + J_{22} + 4J_{12} - 2K_{12} \tag{44}$$

In fact, for a closed shell we need not go to all this trouble to calculate the effect of the two-electron operator, since we can easily write down by inspection the terms that will arise. We will get J_{ii} for the repulsion between the two electrons in each filled MO, $2J_{ij}$ for the repulsion between *each* of the two electrons in the MO ψ_i and the two in every other filled MO ψ_j, corrected by $-K_{ij}$, since one of the two in ψ_j will have its spin parallel to an electron in ψ_i. Thus, summing over the filled MO's

$$\left\langle \ldots \psi_i^\alpha \psi_i^\beta \ldots \psi_j^\alpha \psi_j^\beta \ldots \left| \frac{e^2}{r_{12}} \right| \ldots \psi_i^\alpha \psi_i^\beta \ldots \psi_j^\alpha \psi_j^\beta \ldots \right\rangle$$

$$= \sum_i J_{ii} + 2 \sum_{i<j,j} 2J_{ij} - K_{ij}$$

$$= \sum_i J_{ii} + \sum_{i \neq j, j} 2J_{ij} - K_{ij} \tag{45}$$

$$= \sum_{i,j} 2J_{ij} - K_{ij}$$

where the last equality follows from the fact that $J_{ii} \equiv (\psi_i \psi_i | \psi_i \psi_i) \equiv K_{ii}$.

For the $\psi_2 \to \psi_3$ excited singlet in butadiene,

$$\Psi_V = \frac{1}{\sqrt{2}} (|\psi_1^\alpha \psi_1^\beta \psi_2^\alpha \psi_3^\beta\rangle - |\psi_1^\alpha \psi_1^\beta \psi_2^\beta \psi_3^\alpha\rangle),$$

$$\left\langle \Psi_V \left| \frac{e^2}{r_{12}} \right| \Psi_V \right\rangle = J_{11} + J_{23} + 2J_{12} + 2J_{13} - K_{12} - K_{13} + K_{23} \tag{46}$$

Problem 7.9. Verify Equation (46). Show that each term, except the last, comes from a physical interaction between electrons that can be deduced from inspection of either of the kets.

Because the electron distribution in the paired MO's, ψ_2 and ψ_3, is the same, when we subtract (44) from (46) all the J's cancel and we obtain

$$\left\langle \Psi_V \left| \frac{e^2}{r_{12}} \right| \Psi_V \right\rangle - \left\langle \Psi_N \left| \frac{e^2}{r_{12}} \right| \Psi_N \right\rangle = K_{12} + K_{23} - K_{13} \qquad (47)$$

The K's can be easily evaluated as

$$K_{12} = (\psi_1\psi_2 | \psi_1\psi_2) = \tfrac{1}{16}(\phi_1^2 + \phi_2^2 - \phi_3^2 - \phi_4^2 | \phi_1^2 + \phi_2^2 - \phi_3^2 - \phi_4^2)$$

$$= \tfrac{1}{4}[\gamma_{11} + \gamma_{12} - \gamma_{13} - \tfrac{1}{2}(\gamma_{14} + \gamma_{23})]$$

$$K_{23} = \tfrac{1}{4}[\gamma_{11} - \gamma_{12} + \gamma_{13} - \tfrac{1}{2}(\gamma_{14} + \gamma_{23})] \qquad (48)$$

$$K_{13} = \tfrac{1}{4}[\gamma_{11} - \gamma_{12} - \gamma_{13} + \tfrac{1}{2}(\gamma_{14} + \gamma_{23})]$$

where we have made use of the fact that by symmetry $\gamma_{12} = \gamma_{34}$, $\gamma_{13} = \gamma_{24}$, and $\gamma_{11} = \gamma_{22} = \gamma_{33} = \gamma_{44}$. We can now calculate

$$K_{12} + K_{23} - K_{13} = \tfrac{1}{4}[\gamma_{11} + \gamma_{12} + \gamma_{13} - \tfrac{3}{2}(\gamma_{14} + \gamma_{23})] \qquad (49)$$

The only one of these terms that is different in *cisoid* than in *transoid* butadiene is γ_{14}. Using either (29) or (30), the difference in γ_{14} between the two single bond isomers is on the order of 1 eV, γ_{14} being smaller in the *transoid* isomer. Since γ_{14} is multiplied by $-\tfrac{3}{8}$ in (49), this rough calculation shows that terms arising from the two-electron operator can account for most (about 9 kcal/mole) of the lower energy of the N $\longrightarrow$ V transition in the *cisoid* isomer. Using a more accurate wavefunction for butadiene in the calculation improves the agreement with experiment further.

Problem 7.10.   The wavefunctions for the excited singlets, arising from two different one-electron excitations, $\psi_i \longrightarrow \psi_k$ and $\psi_j \longrightarrow \psi_l$, in the same molecule are:

$$\Psi_1 = \frac{1}{\sqrt{2}} (| \ldots \psi_i^\alpha \psi_k^\beta \psi_j^\alpha \psi_l^\beta \rangle - | \ldots \psi_i^\beta \psi_k^\alpha \psi_j^\alpha \psi_l^\beta \rangle$$

$$\Psi_2 = \frac{1}{\sqrt{2}} (| \ldots \psi_i^\alpha \psi_l^\beta \psi_j^\alpha \psi_l^\beta \rangle - | \ldots \psi_i^\beta \psi_l^\alpha \psi_j^\beta \psi_l^\alpha \rangle$$

Suppose that Ψ_1 and Ψ_2 have the same symmetry. Compute the energy of their interaction, $\langle \Psi_1 | \mathcal{3C} | \Psi_2 \rangle$, in terms of electrostatic repulsion integrals over MO's of the type $(\psi_q \psi_r | \psi_s \psi_t)$. Show that one term in your expression represents the electrostatic interaction between the transition charges, $e\,\psi_i\psi_k$ and $e\,\psi_j\psi_l$. Show, however, that there is another term, not accounted for in the interacting transition dipole model, which is zero only if the two transitions occur in parts of the molecule that are isolated from each other. Is the energy of interaction between the corresponding triplets zero, as the interacting transition dipole model would predict (since the transition dipole for a singlet to triplet excitation is zero)?

Cyclobutadiene

Cyclobutadiene belongs to the group D_{4h}. We can find which representations its MO's span by carrying out the operations of the group on the four p AO's. As in D_{3h}, all the operations in the right-hand half of the character table are similar to those in the left-hand half, but in D_{4h} the latter are obtained from the former by multiplying by the inversion operator. Again, we can restrict our attention to just those in the D_4 subgroup plus the operation of inversion. As we would expect, the u representations in the bottom half of the character table are identical to the g representations in the top half, except for being antisymmetric to inversion. Therefore, the only part of the character table that we really need is

D_{4h}	E	$2C_4$	C_2	$2C_2'$	$2C_2''$	i
A_{1g}	1	1	1	1	1	1 . . .
A_{2g}	1	1	1	−1	−1	1 . . .
B_{1g}	1	−1	1	1	−1	1 . . .
B_{2g}	1	−1	1	−1	1	1 . . .
E_g	2	0	−2	0	0	2 . . .

In operating on the four p AO's to obtain the character (trace) of their transformation matrix, we again need only consider the effect of one operation in each class, since all the others in the class will have the same effect. We find[3]

E	$2C_4$	C_2	$2C_2'$	$2C_2''$	i
4	0	0	0	−2	0

[3] The C_2' axes pass through bonds, while the C_2'' pass through atoms.

From the character table we can see that we will have an a_2, a b_1 and an e representation and that either the e is g and both a_2 and b_1 u or vice versa. To differentiate between the two possibilities we could refer to the complete character table and find the character for another operation. However, we already know the lowest MO is nondegenerate; and because in it all the p orbitals are in phase, it is antisymmetric to inversion. From these two facts we can deduce that a_{2u}, b_{1u} and e_g are the representations spanned by the MO's.

Problem 7.11.   Find the MO's of cyclobutadiene using the D_4 subgroup of D_{4h}. Remember that you must operate on the AO's with every operation in D_4, not just one from each class. Verify that the degenerate MO's *together* transform as e_g.

Let us now determine the symmetries of the states to which we can expect these MO's to give rise. We will have two electrons in the lowest MO of a_{2u} symmetry. The direct product of a_{2u} with itself is A_{1g}; in fact, a closed-shell always belongs to the totally symmetric representation. The symmetries of the electronic states are solely determined by the two electrons in the degenerate orbitals and can be deduced from the fact that

$$e_g \times e_g = A_{1g} + A_{2g} + B_{1g} + B_{2g}$$

It is clear that there are no spatially degenerate states, despite the fact that we began with a pair of degenerate orbitals. Moreover, we can see from the scheme in Figure 7.1 that although there are only four states, there are six distinct ways of putting two electrons into two orbitals while satisfying the Pauli exclusion principle. Therefore, one of the states must necessarily be degenerate. But since none of their spatial wavefunctions are degenerate, it must be the spin function of one of the states that is threefold degenerate; in short, one of the states must be a triplet.

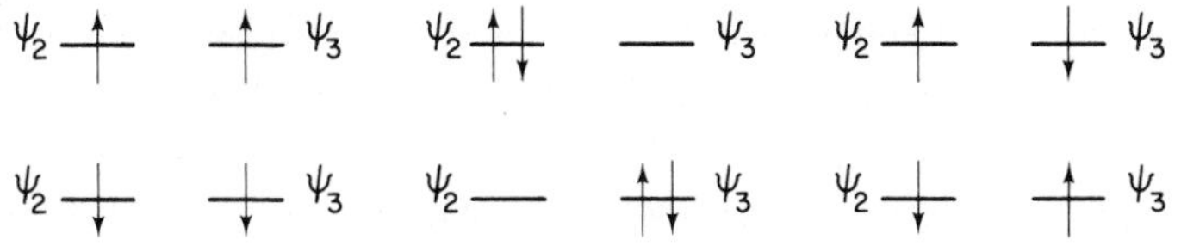

Possible Electronic Configurations in the NBMO's of Cyclobutadiene

Figure 7.1

We can now use the character table to determine the wavefunctions for these states, in much the same way we used it to find MO's as LCAO's, since the states will be linear combinations of kets (electron configurations). Since all the states are g, we again need only employ the first five classes of D_{4h}, which constitute the subgroup D_4. From the behavior of the degenerate MO's, $\psi_2 = \phi_1 - \phi_3$ and $\psi_3 = \phi_2 - \phi_4$, for each operation of D_4, we can determine how two kets formed from them transform, as shown in the following table.

E	$2C_4$	C_2	$2C_2'$	$2C_2''$
ψ_2	$\psi_3, -\psi_3$	$-\psi_2$	$-\psi_3, \psi_3$	$\psi_2, -\psi_2$
ψ_3	$-\psi_2, \psi_2$	$-\psi_3$	$-\psi_2, \psi_2$	$-\psi_3, \psi_3$
$\lvert\psi_2\psi_3\rangle$	$-2\lvert\psi_3\psi_2\rangle$	$\lvert\psi_2\psi_3\rangle$	$2\lvert\psi_3\psi_2\rangle$	$-2\lvert\psi_2\psi_3\rangle$
$\lvert\psi_2\psi_2\rangle$	$2\lvert\psi_3\psi_3\rangle$	$\lvert\psi_2\psi_2\rangle$	$2\lvert\psi_3\psi_3\rangle$	$2\lvert\psi_2\psi_2\rangle$

We can now use the procedure of (19) of Chapter 6 to find each of the states from the transformed kets. Let us begin by generating the state belonging to the A_{1g} representation. If we multiply each of the kets resulting from operating on $\lvert\psi_2\psi_3\rangle$ by $+1$, the character of A_{1g} for each operation, on summing we obtain zero. However, from $\lvert\psi_2\psi_2\rangle$ we get, after normalizing, $\Psi = 1/\sqrt{2}(\lvert\psi_2\psi_2\rangle + \lvert\psi_3\psi_3\rangle)$. Since this ket requires opposite spins for the electrons, this must be an $^1A_{1g}$ state. The characters of A_{2g} when multiplied by the transformed $\lvert\psi_2\psi_3\rangle$ ket give $\lvert\psi_2\psi_3\rangle - \lvert\psi_3\psi_2\rangle$. We have three choices for spin: $\alpha\alpha$, $\beta\beta$ and $\alpha\beta$. The three components of this state are then $\lvert\psi_2^\alpha\psi_3^\alpha\rangle$, $\lvert\psi_2^\beta\psi_3^\beta\rangle$, and $1/\sqrt{2}(\lvert\psi_2^\alpha\psi_3^\beta\rangle - \lvert\psi_3^\alpha\psi_2^\beta\rangle)$; therefore, it is $^3A_{2g}$. From $\lvert\psi_2\psi_3\rangle$ for the characters of B_{1g} we get $\lvert\psi_2\psi_3\rangle + \lvert\psi_3\psi_2\rangle$. Note that for the choice of spin $\alpha\alpha$ this can be rewritten as $\lvert\psi_2^\alpha\psi_3^\alpha\rangle - \lvert\psi_2^\alpha\psi_3^\alpha\rangle$, which vanishes. However, for $\alpha\beta$, $\Psi = (1/\sqrt{2})(\lvert\psi_2^\alpha\psi_3^\beta\rangle + \lvert\psi_3^\alpha\psi_2^\beta\rangle)$, which is the $^1B_{1g}$ state. Finally, for the $^1B_{2g}$ state we find $\Psi = (1/\sqrt{2})(\lvert\psi_2^\alpha\psi_2^\beta\rangle - \lvert\psi_3^\alpha\psi_3^\beta\rangle)$.

Having determined the appropriate wavefunctions for the states,[4] we can proceed to find their relative energies. Because of the intimate relationship between ψ_2 and ψ_3, there is no difference between J_{12} and J_{13} or between K_{12} and K_{13}. Let us, therefore, take as the energy zero the total energy of the molecule, excluding the electrostatic interaction between the

---

[4] Needless to say, this might have been accomplished without the help of group theory by directly computing the mixing between the six configurations.

electrons in the degenerate MO's. Noting that $J_{22} = J_{33}$, the relative energies of the states are then

$$E(^1A_{1g}) = \frac{1}{2}\left\langle \psi_2^\alpha \psi_2^\beta \left| \frac{e^2}{r_{12}} \right| \psi_2^\alpha \psi_2^\beta \right\rangle + \frac{1}{2}\left\langle \psi_3^\alpha \psi_3^\beta \left| \frac{e^2}{r_{12}} \right| \psi_3^\alpha \psi_3^\beta \right\rangle$$

$$+ \left\langle \psi_2^\alpha \psi_2^\beta \left| \frac{e^2}{r_{12}} \right| \psi_3^\alpha \psi_3^\beta \right\rangle = J_{22} + K_{23}$$

$$E(^3A_{2g}) = \left\langle \psi_2^\alpha \psi_3^\alpha \left| \frac{e^2}{r_{12}} \right| \psi_2^\alpha \psi_3^\alpha \right\rangle$$

$$= (\psi_2 \psi_2 | \psi_3 \psi_3) - (\psi_2 \psi_3 | \psi_3 \psi_2) = J_{23} - K_{23}$$

$$E(^1B_{1g}) = \left\langle \psi_2^\alpha \psi_3^\beta \left| \frac{e^2}{r_{12}} \right| \psi_2^\alpha \psi_3^\beta \right\rangle + \left\langle \psi_2^\alpha \psi_3^\beta \left| \frac{e^2}{r_{12}} \right| \psi_3^\alpha \psi_2^\beta \right\rangle$$

$$= J_{23} + K_{23}$$

$$E(^1B_{2g}) = \frac{1}{2}\left\langle \psi_2^\alpha \psi_2^\beta \left| \frac{e^2}{r_{12}} \right| \psi_2^\alpha \psi_2^\beta \right\rangle + \frac{1}{2}\left\langle \psi_3^\alpha \psi_3^\beta \left| \frac{e^2}{r_{12}} \right| \psi_3^\alpha \psi_3^\beta \right\rangle$$

$$- \left\langle \psi_2^\alpha \psi_2^\beta \left| \frac{e^2}{r_{12}} \right| \psi_3^\alpha \psi_3^\beta \right\rangle = J_{22} - K_{23}$$

$$(50)$$

But

$$K_{23} = (\psi_2 \psi_3 | \psi_2 \psi_3) = \tfrac{1}{4}([\phi_1 - \phi_3][\phi_2 - \phi_4] | [\phi_1 - \phi_3][\phi_2 - \phi_4]) \quad (51)$$

which is equal to zero, within the approximation of zero differential overlap. Therefore, the $^3A_{2g}$ and $^1B_{1g}$ states are accidentally degenerate in energy as are $^1A_{1g}$ and $^1B_{2g}$. The former two lie lower than the latter since $J_{22} > J_{23}$. If differential overlap is not neglected, although K_{23} is small, it is positive so that the triplet drops below the lowest singlet. However, if interaction with higher energy configurations is included in order to improve the $^3A_{2g}$ and $^1B_{1g}$ wavefunctions, the singlet is found to be stabilized relative to the triplet. Therefore, even in the most symmetrical molecular geometry, the lowest singlet lies close to, and probably below the triplet state.[5]

The reason for the proximity of these two states is that each has one electron in ψ_2 and one in ψ_3. Since these MO's are confined to separate sets

---

[5] This theoretical result is in excellent agreement with the available experimental evidence. See *J. Amer. Chem. Soc.*, **95**, 8481 (1973) and references therein.

of atoms, there is little difference in electron repulsion energy between the singlet and the triplet that have this electronic configuration—none, in fact, if differential overlap is neglected, since the electron in ψ_2 never appears on the same atom as that in ψ_3, regardless of the spins of these electrons.

The lowest singlet state can be further stabilized, relative to the triplet, by a "pseudo" Jahn-Teller distortion. Such a distortion will not, however, stabilize the singlet by the full $-0.36\,\beta_0$ that we estimated in Chapter 3 for the conversion of square to rectangular cyclobutadiene; for in going to a species consisting of two isolated ethylenes, the electron repulsion in the lowest singlet increases. Since the bonding in singlet cyclobutadiene cannot be maximized while the electron repulsion is simultaneously minimized, the equilibrium geometry for the singlet lies somewhere in between the square and the 1.34×1.52 Å rectangular geometry. Moreover, because the potential energy curve for the distortion is the sum of two functions, one of which (the electron repulsion) increases with distortion while the other (the bonding energy) decreases algebraically, the curve tends to be rather flat and the two minima on it rather shallow.

Problem 7.12.(a) From the character table of D_4 find the states of cyclobutadiene for the choice of degenerate orbitals

$$\psi'_2 = \tfrac{1}{2}(\phi_1 + \phi_2 - \phi_3 - \phi_4)$$

and

$$\psi'_3 = \tfrac{1}{2}(\phi_1 - \phi_2 - \phi_3 + \phi_4)$$

Show again that within the zero differential overlap approximation $^3A_{2g}$ and $^1B_{1g}$ are accidentally degenerate.

(b) To what representation must a vibration belong if it is to mix $^1B_{1g}$ with $^1A_{1g}$? Can you show that a vibration which shortens two opposite bonds and lengthens two others has the requisite symmetry? [Hint: Draw vectors (arrows) on each nucleus that indicate the direction in which each carbon must move in such a mode. Show that the set of vectors has the desired transformation properties.] If the two states were to mix equally, what would the resulting wavefunctions be in terms of the orbitals used for part (a)?

(c) Assume that for one of the two possible distortions from square to 1.34×1.52 Å rectangular cyclobutadiene the energy of the configuration $\Psi_a = |\psi_1^\alpha \psi_1^\beta \psi_2'^\alpha \psi_2'^\beta\rangle$ can be written $E_a = k_\pi x + k_\sigma x^2$ and that of $\Psi_b = |\psi_1^\alpha \psi_1^\beta \psi_3'^\alpha \psi_3'^\beta\rangle$ as $E_b = -k_\pi x + k_\sigma x^2$, where $k_\pi = 0.88\beta_o = -37.8$ kcal/mole and $k_\sigma = -0.52\beta_o = +22.3$ kcal/mole. These are the changes in the sigma and pi energies that we computed in Chapter 3 on going from square $(x = 0)$ to 1.34×1.52 Å rectangular $(x = 1)$ cyclobutadiene with $\beta_o = -43$ kcal/mole. Use (30) to compute the value of the matrix element, K_{ab}, which mixes these two configurations (ignore the small dependence of K_{ab} on x). Write the energy of the lowest singlet state that results from the mixing as a function of x. Show that the minimum occurs at

$$ x = \sqrt{\left(\frac{k_\pi}{2k_\sigma}\right)^2 - \left(\frac{K_{ab}}{k_\pi}\right)^2} $$

and that its energy, relative to that of Ψ_a at $x = 0$, is

$$ E_{\min} = -\frac{k_\pi^2}{4k_\sigma} - \frac{k_\sigma K_{ab}^2}{k_\pi^2} $$

Using the above parameters, by how much does a "pseudo" Jahn-Teller distortion stabilize the lowest singlet of cyclobutadiene in this model calculation? Compare with the Hückel value of $-0.36\beta_0 = 15.5$ kcal/mole.[6]

Problem 7.13.   Petit has observed the reaction shown in Figure 7.2.

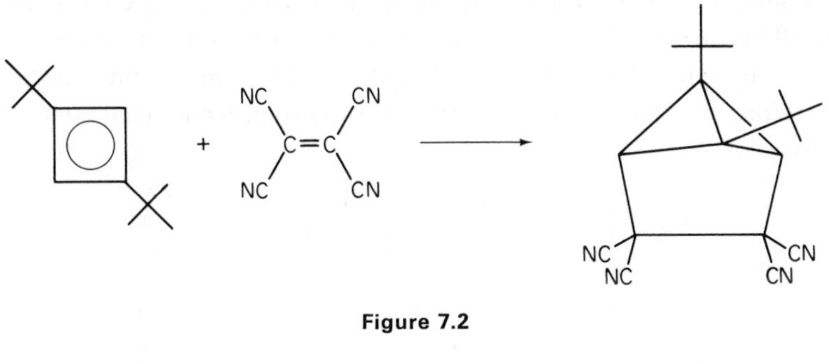

Figure 7.2

(a)   Demonstrate by drawing a correlation diagram that this reaction is allowed to proceed thermally in a concerted fashion if cyclobutadiene is a rectangular singlet.

[6] If the Hückel β_0 is chosen to fit the experimental value of the benzene resonance energy, as suggested in Chapter 3, then the Hückel value for the stabilization of cyclobutadiene by distortion is 11 kcal/mole.

(b) If cyclobutadiene is a ground state triplet, the above reaction must proceed in at least two stages—formation of one or more bonds, followed by spin relaxation to a singlet state before completion of the reaction. Show with a correlation diagram that disrotatory closure of an allylic radical to a cyclopropyl radical is not thermally allowed so that the above reaction cannot proceed in such a fashion. Show also that simultaneous formation of the second and third bonds, as shown in Figure 7.3, is not an allowed reaction.

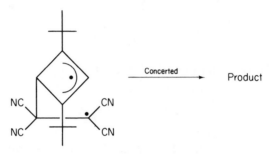

Figure 7.3

Cyclopentadienyl Cation

Cyclopentadienyl belongs to the group D_{5h}; and as usual, we begin by finding the representations spanned by its MO's. Because the π MO's are all constructed from p orbitals, they will all be antisymmetric to the horizontal symmetry plane and belong to doubly primed representations. Therefore, we can consider just the D_5 subgroup, whose character table follows.

D_5	E	$2C_5$	$2C_5^2$	$5C_2$
A_1	1	1	1	1
A_2	1	1	1	-1
E_1	2	$2\cos\frac{2\pi}{5}$	$2\cos\frac{4\pi}{5}$	0
E_2	2	$2\cos\frac{4\pi}{5}$	$2\cos\frac{2\pi}{5}$	0

The representations spanned by the MO's can be determined from the following table, which shows the effect of the operations of the group on the five p AO's.

E	$2C_5$	$2C_5^2$	$5C_2$
5	0	0	-1

Using the fact that $\cos(2\pi/5) = 0.309$ and $\cos(4\pi/5) = -0.809$, these characters reduce to those of $a_2'' + e_1'' + e_2''$. The MO of a_2'' symmetry can rapidly be determined to be

$$\psi_1 = \frac{1}{\sqrt{5}}(\phi_1 + \phi_2 + \phi_3 + \phi_4 + \phi_5)$$

We can also generate the e_1'' MO's, using the D_5 character table. Operating on ϕ_1 and multiplying by the characters of the representation, we obtain

$$\psi_2 = \frac{1}{\sqrt{10}}[2\phi_1 + 0.618(\phi_2 + \phi_5) - 1.618(\phi_3 + \phi_4)]$$

after normalization. We see that this MO has only one node; hence, the MO's of e_1'' symmetry must be the bonding pair. Other functions of this symmetry can be generated by beginning with other atoms in the ring, and it is possible to find a linear combination of them which gives an MO orthogonal to the one of e_1'' symmetry that we have already obtained. However, as was the case with the group D_3 in the previous chapter, we cannot find a set of orthogonal MO's directly from the characters of the degenerate representations in D_5 without going through this process of forming linear combinations of functions generated from the character table. Nevertheless, it is possible to find an orthogonal set *directly* from another character table.

C_5 is a subgroup of D_5, and its character table follows. Beginning with ϕ_1 we can generate two orthogonal MO's from the two sets of characters for the E_1 representation. The complex MO's are[7]

$$\psi_2 = \frac{1}{\sqrt{5}}(\phi_1 + \epsilon\phi_2 + \epsilon^2\phi_3 + \epsilon^{2*}\phi_4 + \epsilon^*\phi_5)$$

---

[7] Equation (19) of Chapter 6 requires that we multiply by the complex conjugates of the characters to obtain the MO's. However, the two sets of complex characters for a degenerate representation are complex conjugates of each other; therefore, using the characters directly from the table to generate the MO's merely changes the arbitrary designation of which we call ψ_2 and which we call ψ_3. Note, however, that with the above designations of ψ_2 and ψ_3, $C_5\psi_2 = \epsilon^*\psi_2$ and $C_5\psi_3 = \epsilon\psi_3$.

and

$$\psi_3 = \frac{1}{\sqrt{5}}(\phi_1 + \epsilon^*\phi_2 + \epsilon^{2*}\phi_3 + \epsilon^2\phi_4 + \epsilon\phi_5)$$

C_5	E	C_5	C_5^2	C_5^3	C_5^4	
A_1	1	1	1	1	1	
E_1	$\begin{cases}1\\1\end{cases}$	$\begin{matrix}\epsilon\\\epsilon^*\end{matrix}$	$\begin{matrix}\epsilon^2\\\epsilon^{2*}\end{matrix}$	$\begin{matrix}\epsilon^{2*}\\\epsilon^2\end{matrix}$	$\begin{matrix}\epsilon^*\\\epsilon\end{matrix}$	$\epsilon = \exp\left(\dfrac{2\pi i}{5}\right)$
E_2	$\begin{cases}1\\1\end{cases}$	$\begin{matrix}\epsilon^2\\\epsilon^{2*}\end{matrix}$	$\begin{matrix}\epsilon^*\\\epsilon\end{matrix}$	$\begin{matrix}\epsilon\\\epsilon^*\end{matrix}$	$\begin{matrix}\epsilon^{2*}\\\epsilon^2\end{matrix}$	

Problem 7.14. Show that these MO's together form a basis for the e_1 representation of D_5. Obtain two orthonormal real orbitals from the complex set.

In addition to allowing us to find rapidly a set of complex orthogonal orbitals, the character table of C_5 tells us how the degenerate e_1 orbitals in complex form transform individually on each of the rotations about the C_5 axis. The fact that each of the complex orbitals is transformed into a multiple of itself on these operations makes these orbitals of the utmost utility in finding the actual states, as we shall see. In contrast, the real orbitals are transformed into linear combinations of each other by these rotations, and this complicates somewhat the determination of the state wavefunctions from the character table.

The representations spanned by the direct product $e_1'' \times e_1''$ are $A_1' + A_2' + E_2'$. We see immediately that, unlike the isoelectronic cyclobutadiene, the cyclopentadienyl cation has a degenerate state, which must be a singlet if the total number of possible configurations is six. In order to find the wavefunctions for the states, we again must first find how the MO's transform under the operations of the group. Since the transformation of the individual complex MO's on rotations about the C_5 axis can be immediately obtained from the C_5 character table, we need only find the effect of the $5C_2$ axes on the MO's. Now the C_2 axis through atom 1 has the effect $\phi_1 \rightarrow -\phi_1$, $\phi_2 \rightarrow -\phi_5$, $\phi_3 \rightarrow -\phi_4$, $\phi_4 \rightarrow -\phi_3$, $\phi_5 \rightarrow -\phi_2$, so

$$C_2^{(1)}\psi_2 = \frac{-1}{\sqrt{5}}(\phi_1 + \epsilon\phi_5 + \epsilon^2\phi_4 + \epsilon^{2*}\phi_3 + \epsilon^*\phi_2) = -\psi_3$$

$$C_2^{(1)}\psi_3 = -\psi_2 \tag{52}$$

For the C_2 axis through atom 2, after ascertaining how the AO's are permuted, we find

$$C_2^{(2)}\psi_2 = \frac{-1}{\sqrt{5}}(\phi_3 + \epsilon\phi_2 + \epsilon^2\phi_1 + \epsilon^{2*}\phi_5 + \epsilon^*\phi_4) = -\epsilon^2\psi_3$$

$$C_2^{(2)}\psi_3 = -\epsilon^{2*}\psi_2 \tag{53}$$

We can similarly deduce that the other three C_2 axes take ψ_2 into $-\epsilon^*$, $-\epsilon$, and $-\epsilon^{2*}$ times ψ_3, while ψ_3 is respectively transformed into $-\epsilon$, $-\epsilon^*$, and $-\epsilon^2$ times ψ_2. The effect of the $5C_2$ axes on the two kets $|\psi_2\psi_3\rangle$ and $|\psi_2\psi_2\rangle$ is then

$$\sum_{i=1}^{5} C_2^{(i)}|\psi_2\psi_3\rangle = (1 + 2\epsilon\epsilon^* + 2\epsilon^2\epsilon^{2*})|\psi_3\psi_2\rangle = 5|\psi_3\psi_2\rangle$$

$$\sum_{i=1}^{5} C_2^{(i)}|\psi_2\psi_2\rangle = [1 + (-\epsilon^2)^2 + (-\epsilon^*)^2 + (-\epsilon)^2 + (-\epsilon^{2*})^2]|\psi_3\psi_3\rangle$$

$$= [1 + \epsilon^* + \epsilon^{2*} + \epsilon^2 + \epsilon]|\psi_3\psi_3\rangle = 0 \tag{54}$$

A summary of the effect of all the operations of the group D_5 on these kets follows.

E	$2C_5$	$2C_5^2$	$5C_2$
$|\psi_2\psi_3\rangle$	$2|\psi_2\psi_3\rangle$	$2|\psi_2\psi_3\rangle$	$5|\psi_3\psi_2\rangle$
$|\psi_2\psi_2\rangle$	$2\cos\dfrac{4\pi}{5}|\psi_2\psi_2\rangle$	$2\cos\dfrac{2\pi}{5}|\psi_2\psi_2\rangle$	0

From the characters of the A_1 representation we obtain $|\psi_2\psi_3\rangle + |\psi_3\psi_2\rangle$. The spin function $\alpha\alpha$ causes the wavefunction to vanish so $(1/\sqrt{2})\cdot$ $(|\psi_2^\alpha\psi_3^\beta\rangle + |\psi_3^\alpha\psi_2^\beta\rangle)$ is $^1A_1'$. Multiplying the transformed kets by the characters of the A_2' representation and summing, the choice $\alpha\alpha$ gives $|\psi_2^\alpha\psi_3^\alpha\rangle$, and $\alpha\beta$ gives $(1/\sqrt{2})(|\psi_2^\alpha\psi_3^\beta\rangle - |\psi_3^\alpha\psi_2^\beta\rangle)$, so these are components of $^3A_2'$. Finally, we see that although the characters of A_1 and A_2 give zero when multiplied by the transformed $|\psi_2\psi_2\rangle$ kets, the characters of E_2 give $|\psi_2^\alpha\psi_2^\beta\rangle$ as one component of the degenerate $^1E_2'$ representation, $|\psi_3^\alpha\psi_3^\beta\rangle$ being the other.

Problem 7.15. Show that $|\psi_2^\alpha\psi_2^\beta\rangle$ and $|\psi_3^\alpha\psi_3^\beta\rangle$ together form a basis for the E_2 representation of D_5 by showing how they transform for each class of operations.

We can now compute the relative energies of the states. For this computation, we could, if we wish, transform to a set of real orbitals. Since we can write a set of orthogonal real orbitals as linear combinations of the complex ones, namely

$$\psi'_2 = \frac{1}{\sqrt{2}}(\psi_2 + \psi_3) \quad \text{and} \quad \psi'_3 = \frac{i}{\sqrt{2}}(\psi_2 - \psi_3)$$

we can always express the complex orbitals as linear combinations of the real orbitals. Thus:

$$\psi_2 = \frac{1}{\sqrt{2}}(\psi'_2 - i\psi'_3) \quad \text{and} \quad \psi_3 = \frac{1}{\sqrt{2}}(\psi'_2 + i\psi'_3)$$

so that we can write the states in terms of the real (primed) orbitals. Nevertheless, there is no need to carry out this transformation, since we can compute the energies directly from the complex orbitals. We must be careful to remember, however, that the wavefunctions in our bras are all complex conjugates. This follows from the fact that the integral $\int \psi_i^* G \psi_j \, d\tau$ in bra-ket notation becomes $\langle \psi_i | G | \psi_j \rangle$, so $\langle \psi_i |$ is really ψ_i^*.

The complex orbitals have an advantage over the real ones in computing the energy. Because the electron density is evenly distributed over all the atoms in both degenerate complex orbitals, $J_{22} = J_{23} = J_{33}$. Therefore, for calculating relative energies, we need only compute the exchange integral, K_{23}. The $^3A'_2$ state lies K_{23} below the degenerate $^1E'_2$ which lies K_{23} below $^1A'_1$, where K_{23} is given by

$$K_{23} = \left\langle \psi_2^\alpha \psi_3^\beta \left| \frac{e^2}{r_{12}} \right| \psi_3^\alpha \psi_2^\beta \right\rangle = (\psi_2^* \psi_3 | \psi_3^* \psi_2) = (\psi_3^2 | \psi_2^2) \qquad (55)$$

The last equality in (55) follows from the fact that the degenerate orbitals are each other's complex conjugates. Using identities like

$$\epsilon^2 \times \epsilon^2 = \epsilon^4 \equiv \epsilon^*$$

and

$$\epsilon^{2*} \times \epsilon^{2*} \equiv \epsilon^3 \times \epsilon^3 = \epsilon^6 \equiv \epsilon^5 \times \epsilon = \epsilon$$

and computing the required repulsion integrals between AO's from Equation (30), we find

$$(\psi_3^2 \mid \psi_2^2) = \frac{1}{25}(\phi_1^2 + \epsilon^{2*}\phi_2^2 + \epsilon\phi_3^2 + \epsilon^*\phi_4^2 + \epsilon^2\phi_5 \mid \phi_1^2$$
$$+ \epsilon^2\phi_2^2 + \epsilon^*\phi_3^2 + \epsilon\phi_4^2 + \epsilon^{2*}\phi_5^2)$$

$$= \frac{1}{5}\left(\gamma_{11} + 2\cos\frac{4\pi}{5}\gamma_{12} + 2\cos\frac{2\pi}{5}\gamma_{13}\right) \tag{56}$$

$$= \frac{1}{5}(11.1 - 1.62 \times 7.5 + 0.62 \times 5.2)\,\mathrm{eV}$$

$$= 0.6\,\mathrm{eV}$$

Thus, the triplet, unlike that in cyclobutadiene, is calculated to lie 0.6 eV below the lowest singlet in the most symmetrical configuration of the cyclopentadienyl cation. This does not assure us that a Jahn-Teller distortion will not make the actual ground state a distorted singlet, but it does suggest that the triplet is much more likely to be the ground state or close to it in the cyclopentadienyl cation than in the isoelectronic cyclobutadiene. Breslow and his co-workers have, in fact, experimentally found the triplet to be the ground state of the cyclopentadienyl cation.

The reason that the triplet state lies well below the lowest singlet in the symmetrical cyclopentadienyl cation is that, unlike the case in the cyclobutadiene ring, it is impossible to find two degenerate orbitals that have no atoms in common. Therefore, the triplet lies below the lowest singlet because it keeps the two electrons in the degenerate MO's from simultaneously being in the same AO.

Problem 7.16.   Why are the two states which belong to the $^1E_2'$ representation not split by the matrix element $\langle \psi_2^x \psi_2^\beta \mid e^2/r_{12} \mid \psi_3^x \psi_3^\beta \rangle$? [Hint: Calculate it.]

A 4n Atom Rule for Singlet Ground States

We have seen that the two $4n$ π electron systems, cyclobutadiene and cyclopentadienyl cation, differ markedly in that the former is predicted to have a singlet state that lies very close to the triplet in the most symmetrical molecular configuration, while in the cyclopentadienyl cation, the triplet lies well below the singlet in the molecular geometry of the highest symmetry. Thus, it is far more likely in the former molecule than in the latter that the singlet will be the ground state, and this prediction seems to be in excellent agreement with experiment.

We are now led to inquire whether this difference may generally be expected of $4n$ π electron systems in rings of $4n$ and $4n \pm 1$ atoms. We can begin to investigate this question through the use of group theory. A planar symmetrical ring of m atoms belongs to the point group D_{mh}. If the molecule has $4n$ π electrons, two electrons must be placed in two orbitals which belong to the degenerate e''_n ($m = 4n \pm 1$) or e_{ng} ($m = 4n$) representation of the group. The symmetries of the electronic states so formed may be deduced from the representations spanned by the direct product $e''_n \times e''_n$ in a ring of $4n \pm 1$ atoms and $e_{ng} \times e_{ng}$ in a ring of $4n$ atoms. The former gives rise to states A'_1, $E'_{2n-(1\pm1)/2}$, and A'_2 in a $4n \pm 1$ membered ring. The A'_2 state is the triplet which lies $2K_{ij}$ lower than the $^1A'_1$ state and K_{ij} lower than the pair of degenerate singlet states, where ψ_i and ψ_j are the degenerate *complex* molecular orbitals and

$$K_{ij} = \left\langle \psi_i^\alpha \psi_j^\beta \left| \frac{e^2}{r_{12}} \right| \psi_j^\alpha \psi_i^\beta \right\rangle = (\psi_i^* \psi_j | \psi_j^* \psi_i) \tag{57}$$

In the groups D_{mh} ($m = 4n$) the direct product $e_{ng} \times e_{ng}$ spans A_{1g}, B_{1g}, B_{2g}, and A_{2g}. Again $^1A_{1g}$ and $^3A_{2g}$ are split by $2K_{ij}$, but the other two singlet states are, in general, no longer degenerate. They are split by $2K'_{ij}$, where

$$K'_{ij} = \left\langle \psi_i^\alpha \psi_i^\beta \left| \frac{e^2}{r_{12}} \right| \psi_j^\alpha \psi_j^\beta \right\rangle = (\psi_i^* \psi_j | \psi_i^* \psi_j) \tag{58}$$

The fact that group theory shows a degenerate pair of states in rings of $4n \pm 1$ atoms means that in these systems the symmetry of the degenerate orbitals must cause K'_{ij} to vanish. However, for rings of $4n$ atoms symmetry does not cause K'_{ij} to vanish. In fact, in these systems it can be shown that $K'_{ij} = K_{ij}$, if the usual approximation of zero differential overlap is made. This means that, within the validity of this approximation, in $4n$ π electron systems consisting of $4n$ atoms the lowest singlet and triplet have the same energy, since $J_{ii} = J_{ij}$ when complex MO's are used.

In order that $K'_{ij} = K_{ij}$, comparison of Equations (57) and (58) shows that a sufficient condition is

$$\psi_j^* \psi_i = \psi_i^* \psi_j \tag{59}$$

This equality can be demonstrated using two special properties of ψ_i and ψ_j. The first is that because they are a degenerate pair, they can be written as complex conjugates of each other:

$$\psi_i = \psi_j^*$$ (60)

The second is that as two nonbonding molecular orbitals in an alternant hydrocarbon, if ψ_i is written

$$\psi_i = \sum_s^{2n} c_s\phi_s + \sum_u^{2n} c_u\phi_u$$ (61)

where the atomic orbitals have been divided into two sets, s and u, that alternate around the ring,[8] then, using the pairing theorem, ψ_j may be written

$$\psi_j = \sum_s^{2n} c_s\phi_s - \sum_u^{2n} c_u\phi_u$$ (62)

Substituting (61) and (62) into (60) yields

$$\sum_s c_s\phi_s + \sum_u c_u\phi_u = \sum_s c_s^*\phi_s - \sum_u c_u^*\phi_u$$ (63)

Equating coefficients,

$$c_s = c_s^* \quad \text{and} \quad c_u = -c_u^*$$ (64)

showing that the c_s must be pure real and the c_u pure imaginary numbers.
Use of (61), (62), and (64) gives

$$\psi_j^*\psi_i = (\sum_s c_s\phi_s + \sum_u c_u\phi_u)^2$$

and (65)

$$\psi_i^*\psi_j = (\sum_s c_s\phi_s - \sum_u c_u\phi_u)^2$$

Finally, since zero differential overlap implies $\phi_s\phi_u = 0$, the cross terms vanish and

[8] These are the usual starred and unstarred sets, but we refrain from starring in order to avoid confusion with the sign for a complex conjugate, which is also a star.

$$\psi_j^* \psi_i = (\sum_s c_s \phi_s)^2 + (\sum_u c_u \phi_u)^2$$

$$= (\sum_s c_s \phi_s)^2 + (-\sum_u c_u \phi_u)^2 = \psi_i^* \psi_j \qquad (66)$$

The different situations with regard to the energy of the states in $4n \pm 1$ and $4n$ systems are summarized in Figure 7.4.

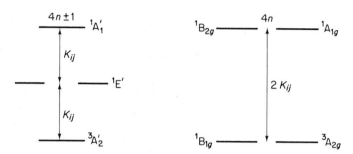

Energies of Low–Lying States in Rings of $4n \pm 1$ and $4n$ Atoms Containing $4n$ Electrons

Figure 7.4

In fully symmetrical rings of $4n \pm 1$ atoms, the triplet lies lowest. Even if the ground state of the system is a distorted singlet, the triplet, once populated, might show some stability if the distortions caused by the molecular vibrations are sufficiently small. In contrast, even in fully symmetrical rings of $4n$ atoms the lowest singlet lies very close to the triplet, and a pseudo Jahn-Teller distortion is capable of further lowering the energy of the former. Therefore, it is predicted that the triplet state should play a much less important role in rings of $4n$ atoms than in those consisting of $4n \pm 1$ atoms. Since as ring size increases, the magnitude of K_{ij} decreases, the stability of the triplet relative to the undistorted singlet in $4n \pm 1$ atom rings should be greatest in small systems.

Group theory shows that rings of $4n \pm 2$ atoms containing $4n$ electrons also have a degenerate pair of singlet states. Therefore, they are expected, like the isoelectronic $4n \pm 1$ atom rings, to have a symmetrical triplet state of some potential stability. Hoijtink has observed stable triplet states in the dianions of triphenylbenzene and decacylene; and more recently, Wasserman has found the dipositive ion of hexachlorobenzene to be a ground state triplet.[9]

---

[9] See *J. Amer. Chem, Soc.* **96,** 1965 (1974) for a report of this work and leading references to other epr studies of triplet ground state molecules.

Problem 7.17. Use (61), (62), and (64) to find two real orbitals, each of which is paired with itself. Write these orbitals as linear combinations of the complex orbitals and vice versa. Substitute for the complex orbitals in (57) and (58) the appropriate linear combinations of the real MO's. Show that, with neglect of differential overlap, $K_{ij} = K'_{ij}$ because the two degenerate real orbitals have no atoms in common. Explain in words why the lowest singlet and the triplet have the same energy in symmetrical rings of $4n$ atoms.

Problem 7.18. Calculate the total energy of the lowest singlet and triplet states of trimethylenemethane (See Problem 6.13). Do the same for the conformation with one CH_2 group twisted out of conjugation. For the sake of simplicity, use Equation (30) to calculate all γ's and take $\beta = -1.5$ eV. Do your calculations agree with the experimental result that, despite the fact that the Hückel energy of the wavefunction for fully delocalized trimethylenemethane is $-2(\sqrt{3} - \sqrt{2})\beta$ lower than that of the wavefunction for allyl plus an electron localized in a p orbital, the former is preferred only for the triplet?

Excited States of Benzene—The "Hole" Formalism

The problem of computing the energies of the lowest excited states of benzene relative to that of the ground state is similar to the calculations on molecules with open-shell ground states that we have already discussed. The electron that is excited in benzene goes into an orbital of e_{2u} symmetry; and to determine the representations spanned by the excited states, we must find the direct product of e_{2u} with the representation to which the wavefunction for the remaining three electrons in the e_{1g} orbitals belongs. We need not worry about the two electrons in the lowest benzene MO, for they form a closed-shell, and we know that a closed-shell always belongs to the totally symmetric representation. Now we might proceed to find the representation spanned by the three electrons remaining in the degenerate bonding MO's by computing the direct product $e_{1g} \times e_{1g} \times e_{1g}$. If we were to carry out the work necessary to compute this direct product, however, we would discover that it spans several different representations. Moreover, not all of them correspond to the representation to which the actual wavefunction belongs. The reason that we obtain extraneous representations is that some of them correspond to wavefunctions that violate the exclusion principle. We could, of course, just write down for the three electrons wavefunctions that obey the exclusion principle; and, by discovering how they transform, we could deduce the representation(s) to which they belong. However, there is an easier way to proceed.

We know that four electrons in the e_{1g} orbitals constitute a closed-shell, whose wavefunction consequently belongs to the totally symmetric representation. In order to give the totally symmetric representation, the representation to which the wavefunction for three electrons belongs must be the same as that to which the wavefunction for one electron in this MO belongs. The wavefunction for one electron in e_{1g}, of course, belongs to e_{1g}; thus, this is the representation to which the wavefunction for the three remaining electrons must also belong. As illustrated by this example, when an unfilled shell has more than half its complement of electrons, it is often easier to work with the symmetry of the missing electrons, or electron "holes" in the closed-shell, than to deal directly with the transformation properties of those present.

Problem 7.19. Use the part of the character table for D_{6h}, found in the appendix along with that for C_6, to deduce the possible symmetries of the lowest excited states of benzene. To which is a transition allowed? What vibrational modes can mix some of the wavefunction for the allowed transition into the "forbidden" ones, thus making the latter partially allowed? Use the character table to generate the wavefunctions for the excited states, remembering that *each* symmetry type can either be a singlet or triplet. [Hint: To generate the wavefunctions for the excited states, you will have to determine how kets like $|\psi_2\psi_2\psi_3\psi_4\rangle$ and $|\psi_2\psi_2\psi_3\psi_5\rangle$ transform under the operations of the group. Note, however, that the closed-shell $|\psi_2\psi_2\psi_3\psi_3\rangle$ goes into itself on every operation of the group, and use this fact to establish the relationship between the transformation properties of $|\psi_2\psi_2\psi_3\rangle$ and $|\psi_3\rangle$. You may find it convenient to use this relationship between $|\psi_2\psi_2\psi_3\rangle$ and a $|\psi_3\rangle$ "hole" in the closed-shell, $|\psi_2\psi_2\psi_3\psi_3\rangle$, to determine how $|\psi_2\psi_2\psi_3\rangle$ transforms.]

Problem 7.20. The allowed transition in benzene occurs at 6.76 eV. Assume that it is valid to use (11) of Chapter 3 to correct the spectroscopic β in ethylene to the benzene bond length. Use the spectroscopic β for benzene so found, together with the repulsion integrals calculated from (30), to compute a theoretical estimate of the energy of the allowed transition in benzene.

Summary

In this chapter we reexamined, with the inclusion of electron repulsion in the Hamiltonian, some of the molecules we had already treated using Hückel theory. In order that the inclusion of the two-electron repulsion operator should not make our calculations too unwieldy, we were forced to adopt the formal neglect of differential overlap. We saw that this approximation is related to the Mulliken approximation to an overlap distribution, from which the zero differential overlap approximation arises naturally. In order that our electron repulsion integrals between AO's would allow us to accurately calculate experimentally determined energies, we found it necessary, especially for one-center integrals, to adopt semiempirical values to allow for correlation and other effects.

Applying this semiempirical method to ethylene, we saw how the energy difference between the singlet and triplet excited states arises naturally from their different wavefunctions, the triplet being covalent and the singlet ionic in nature. We also found that the doubly excited configuration is mixed into the ground state wavefunction to reduce the ionic character in the latter. The improvement of the ground state wavefunction through configuration interaction is especially important when the amount of bonding is small; or, to state it more generally, configuration interaction becomes increasingly important as two configurations approach degeneracy. In discussing configuration interaction in ethylene, we observed that eigenvalue problems can be formulated conveniently in matrix notation, a point we shall explore more fully in the next chapter.

In butadiene we saw how a net electrostatic attraction between the termini in the lowest singlet excited state gives rise naturally to an energy difference between *cisoid* and *transoid* molecules for the $N \rightarrow V$ transition.

Turning to cyclobutadiene, we used group theory extensively to find the wavefunctions for the states of this open-shell molecule. Our calculation showed a singlet state to be accidentally degenerate with the triplet, even in the molecular geometry of highest symmetry, suggesting that cyclobutadiene is a ground state singlet.

In contrast, in the isoelectronic cyclopentadienyl cation the triplet was calculated to lie below the degenerate singlet in the most symmetrical geometry. We were able to show generally, using group theory, that the triplet is most likely to be the ground state in systems containing $4n$ electrons in rings of $4n \pm 1$ or $4n \pm 2$ atoms; whereas, in rings of $4n$ atoms the lowest singlet lies very close to the triplet, even in the most symmetrical molecular geometry.

APPENDIX FOR CHAPTER 7

D_{6h}	E	$2C_6$	$2C_3$	C_2	$3C_2$	$3C_2''$	i ...
A_{1g}	1	1	1	1	1	1	1 ...
A_{2g}	1	1	1	1	-1	-1	1 ...
B_{1g}	1	-1	1	-1	1	-1	1 ...
B_{2g}	1	-1	1	-1	-1	1	1 ...
E_{1g}	2	1	-1	-2	0	0	2 ...
E_{2g}	2	-1	-1	2	0	0	2 ...

(x,y) transform together as E_{1u} and z transforms as A_{2u} in D_{6h}

C_6	E	C_6	C_3	C_2	C_3^2	C_6^5
A	1	1	1	1	1	1
B	1	-1	1	-1	1	-1
E_1	$\left\{\begin{array}{l}1\\1\end{array}\right.$	$\begin{array}{c}\epsilon\\\epsilon^*\end{array}$	$\begin{array}{c}-\epsilon^*\\-\epsilon\end{array}$	$\begin{array}{c}-1\\-1\end{array}$	$\begin{array}{c}-\epsilon\\-\epsilon^*\end{array}$	$\left.\begin{array}{c}\epsilon^*\\\epsilon\end{array}\right\}$
E_2	$\left\{\begin{array}{l}1\\1\end{array}\right.$	$\begin{array}{c}-\epsilon^*\\-\epsilon\end{array}$	$\begin{array}{c}-\epsilon\\-\epsilon^*\end{array}$	$\begin{array}{c}1\\1\end{array}$	$\begin{array}{c}-\epsilon^*\\-\epsilon\end{array}$	$\left.\begin{array}{c}-\epsilon\\-\epsilon^*\end{array}\right\}$

$\epsilon = e^{2\pi i/6}$

FURTHER READING

Robert G. Parr, *The Quantum Theory of Molecular Electronic Structure*, W. A. Benjamin, New York, 1963, gives an excellent account of the semiempirical π electron theory which he helped develop. The book contains numerous references to the literature as well as selected reprints of original papers.

8

Molecular Orbitals from
Self-Consistent Field Theory

In the preceding chapter we saw how to calculate total energies when
the electron repulsion operator is included in the Hamiltonian. We noted
that the presence of a two-electron operator in the true Hamiltonian
makes it impossible to use the Schrödinger equation to calculate one-
electron wavefunctions, since the wavefunction of each electron depends
on that of every other. We might be able to use in our calculations approximate one-electron wavefunctions, obtained perhaps from HMO theory,
if we were prepared to carry out extensive configuration interaction.
However, even though we could get good wavefunctions by this procedure,
there are major drawbacks to having to write all our wavefunctions as
linear combinations of Slater determinants, when it might be possible to
find a single Slater determinant that would represent the wavefunction
for a state very well. In this chapter we will see how to obtain the best
one-electron wavefunctions (MO's), written as LCAO's, with which to
make up a Slater determinant.

The Variational Principle

In Hückel theory we were able to find our MO's merely by looking for LCAO's that would be eigenfunctions of our one-electron Hamiltonian. We were also able to show that the wavefunctions so found minimized the energy. It is physically reasonable that the wavefunctions we calculate should be chosen so that they minimize the energy of a system, since the true wavefunctions will be those, which for each state, allow the system to have the lowest possible energy. Because the wavefunctions that we compute will only be approximations to the actual ones, by ensuring that the former minimize the calculated total energy, we can be certain that the true energy will be a lower bound to the energy of our approximate wavefunctions.

We can prove this very easily. Suppose we try to calculate the ground state wavefunction, Ψ_0, of a molecule, and that we obtain the wavefunction Ψ_0'. Assuming that the true wavefunctions form a complete set, we can expand our approximate wavefunction as a linear combination of them.

$$\Psi_0' = \sum_{n=0} c_n \Psi_n \tag{1}$$

Now let us calculate the energy of our approximate wavefunction

$$\langle \Psi_0' | \mathcal{3C} | \Psi_0' \rangle = E_0' = \sum_{m,n} \langle c_m \Psi_m | \mathcal{3C} | c_n \Psi_n \rangle = \sum_{n=0} \langle c_n | c_n \rangle E_n \tag{2}$$

The last equality comes from the fact that each of the Ψ_n satisfies

$$\mathcal{3C} | \Psi_n \rangle = E_n | \Psi_n \rangle \tag{3}$$

and the assumption that the Ψ_n have been chosen so that they form an orthonormal set. Subtracting E_0, the true energy of the ground state wavefunction, from both sides of (2) gives

$$E_0' - E_0 = \sum_{n=0} \langle c_n | c_n \rangle E_n - E_0 = \sum_{n=0} \langle c_n | c_n \rangle (E_n - E_0) \tag{4}$$

The last step follows from the fact that $\sum \langle c_n | c_n \rangle = 1$. Since each term, $\langle c_n | c_n \rangle$, in the sum is positive, as is $E_n - E_0$, the right-hand side of (4) is positive unless, except for c_0, all the $c_n = 0$. Therefore, $E'_0 \geq E_0$, so that the energy of an approximate wavefunction is, indeed, greater than that of the true one.

Self-Consistent Field (SCF) Theory

In the previous chapter we found the energy of a closed-shell ground state $|\Psi_0\rangle = |\psi_1^\alpha \psi_1^\beta \ldots \psi_n^\alpha \psi_n^\beta\rangle$ to be

$$E_0 = \sum_{A>B,B} \frac{Z'_A Z'_B e^2}{r_{AB}} + 2\sum_i h_i + \sum_{i,j} 2J_{ij} - K_{ij} \tag{5}$$

with the last two sums over just the filled MO's and the definitions

$$h_i = \langle \psi_i | \hbar | \psi_i \rangle \tag{6}$$

$$J_{ij} = (\psi_i \psi_i | \psi_j \psi_j) \tag{7}$$

$$K_{ij} = (\psi_i \psi_j | \psi_j \psi_i) \tag{8}$$

If we wish to minimize the energy with respect to variations of the ψ_i, we must be careful to include the constraint that these one-electron wavefunctions remain orthonormal. We can do this by the Lagrange method of undetermined multipliers. We require that

$$S_{ij} = \langle \psi_i | \psi_j \rangle = \delta_{ij} \tag{9}$$

so we add $-2\sum \epsilon_{ij} S_{ij}$ to (5), where the ϵ_{ij} are constants to be determined, and we minimize the resulting function with respect to small changes, δ, in the wavefunction by requiring

$$2\sum_i \delta h_i + \sum_{i,j} (2\,\delta J_{ij} - \delta K_{ij} - 2\epsilon_{ij}\,\delta S_{ij}) = 0 \tag{10}$$

Two of the terms in (10) can be obtained directly from the definitions (6) and (9).

$$\delta h_i = \langle \delta \psi_i | \hbar | \psi_i \rangle + \langle \psi_i | \hbar | \delta \psi_i \rangle \tag{11}$$

$$\delta S_{ij} = \langle \delta \psi_i | \psi_j \rangle + \langle \psi_i | \delta \psi_j \rangle \tag{12}$$

If we now implicitly also define two new one-electron operators by the equations

$$J_{ij} \equiv \langle \psi_i | \mathcal{J}_j | \psi_i \rangle = \langle \psi_j | \mathcal{J}_i | \psi_j \rangle \tag{13}$$

$$K_{ij} \equiv \langle \psi_i | \mathcal{K}_j | \psi_i \rangle = \langle \psi_j | \mathcal{K}_i | \psi_j \rangle \tag{14}$$

we can compute

$$
\begin{aligned}
\delta(2J_{ij} - K_{ij}) &= \langle \delta\psi_i | 2\mathcal{J}_j - \mathcal{K}_j | \psi_i \rangle + \langle \delta\psi_j | 2\mathcal{J}_i - \mathcal{K}_i | \psi_j \rangle \\
&\quad + \langle \psi_i | 2\mathcal{J}_j - \mathcal{K}_j | \delta\psi_i \rangle + \langle \psi_j | 2\mathcal{J}_i - \mathcal{K}_i | \delta\psi_j \rangle
\end{aligned}
\tag{15}
$$

Since we are summing over two dummies, i and j, in (10)

$$\sum_{i,j} \langle \psi_i | 2\mathcal{J}_j - \mathcal{K}_j | \delta\psi_i \rangle = \sum_{i,j} \langle \psi_j | 2\mathcal{J}_i - \mathcal{K}_i | \delta\psi_j \rangle \tag{16}$$

$$\sum_{i,j} \langle \psi_i | \delta\psi_j \rangle \epsilon_{ij} = \sum_{i,j} \langle \psi_j | \delta\psi_i \rangle \epsilon_{ji} \tag{17}$$

Thus, Equation (10) becomes

$$
\begin{aligned}
&\sum_i \langle \delta\psi_i | 2\hbar + 2\sum_j 2\mathcal{J}_j - \mathcal{K}_j | \psi_i \rangle - 2 \sum_{i,j} \langle \delta\psi_i | \psi_j \rangle \epsilon_{ij} \\
&+ \sum_i \langle \psi_i | 2\hbar + 2\sum_j 2\mathcal{J}_j - \mathcal{K}_j | \delta\psi_i \rangle - 2 \sum_{i,j} \langle \psi_j | \delta\psi_i \rangle \epsilon_{ji} = 0
\end{aligned}
\tag{18}
$$

Hermiticity in bra-ket notation is defined as

$$\langle \Gamma_i | G | \Gamma_j \rangle = \int \Gamma_i^* G \Gamma_j = \int \Gamma_j G^* \Gamma_i^* = \langle \Gamma_j^* | G^* | \Gamma_i^* \rangle = \langle \Gamma_j | G | \Gamma_i \rangle^* \tag{19}$$

so (18) can be rewritten

$$
\begin{aligned}
&\sum_i \langle \delta\psi_i | 2\hbar + 2\sum_j 2\mathcal{J}_j - \mathcal{K}_j | \psi_i \rangle - 2 \sum_{i,j} \langle \delta\psi_i | \psi_j \rangle \epsilon_{ij} \\
&+ \sum_i \langle \delta\psi_i^* | 2\hbar^* + 2\sum_j 2\mathcal{J}_j^* - \mathcal{K}_j^* | \psi_i^* \rangle \\
&- 2 \sum_{i,j} \langle \delta\psi_i^* | \psi_j^* \rangle \epsilon_{ji} = 0
\end{aligned}
\tag{20}
$$

The coefficients of $\delta\psi_i$ and $\delta\psi_i^*$ must vanish independently so

$$\left| \hbar + \sum_j 2\mathfrak{J}_j - \mathfrak{K}_j \left| \psi_i \right\rangle = \sum_j \epsilon_{ij} \left| \psi_j \right\rangle \right.$$

$$\left| \hbar^* + \sum_j 2\mathfrak{J}_j^* - \mathfrak{K}_j^* \left| \psi_i^* \right\rangle = \sum_j \epsilon_{ji} \left| \psi_j^* \right\rangle \right. \tag{21}$$

The complex conjugate of the left-hand side of the second equation is equal to the left-hand side of the first, so that by taking the complex conjugate of the second and subtracting, we obtain

$$\sum_j \left| \psi_j \right\rangle (\epsilon_{ij} - \epsilon_{ji}^*) = 0 \tag{22}$$

Equation (22) implies, since the ψ_j are linearly independent, that $\epsilon_{ij} = \epsilon_{ji}^*$. This shows that the matrix $[\epsilon_{ij}]$ must be Hermitian, which guarantees that we can diagonalize it to find a set of real eigenvalues. Thus, after diagonalizing $[\epsilon_{ij}]$, (21) becomes

$$\mathfrak{F} \left| \psi_i \right\rangle \equiv \left| \hbar + \sum_j 2\mathfrak{J}_j - \mathfrak{K}_j \left| \psi_i \right\rangle = \epsilon_i \left| \psi_i \right\rangle \right. \tag{23}$$

where $\mathfrak{F}$ is known as the Fock operator; and Equation (23) is the Hartree-Fock equation for finding the one-electron wavefunctions, ψ_i, which give the minimum energy that can be obtained with a single Slater determinant. It can be shown (Problem 8.1) that if the eigenfunctions of $\mathfrak{F}$ are used to carry out a configuration interaction calculation to improve the ground state wavefunction, singly excited states will not mix with the ground state. This result, known as Brillouin's theorem, implies that because doubly excited configurations usually lie well above the ground state in energy (and hence will only mix a small amount with the ground state), configuration interaction will not lower the energy of the SCF ground state wavefunction to any great extent. This is the case unless a doubly excited configuration is close to the lowest one in energy or unless one mixes in a very large number of configurations.

One might wonder why there is a certain arbitrariness in the ψ_i, since it appears from (21) and (23) that they either may or may not be eigenfunctions of $\mathfrak{F}$ and still minimize the energy of the many-electron wavefunction for the ground state. This should not surprise us, however, for we have seen that we may always add multiples of rows in a Slater determinant without changing its value; therefore, there are really an infinite number of possible combinations of one-electron wavefunctions within the determinant that will minimize the energy. Nevertheless, we will in general wish to find the set that are eigenfunctions of $\mathcal{F}$. Here again symmetry can help us; for, since symmetry elements commute with $\mathcal{F}$, the eigenfunctions of the symmetry operators are also eigenfunctions, or linear combinations of eigenfunctions, of $\mathcal{F}$. Moreover, wavefunctions that belong to different representations of the molecular symmetry group will not be mixed by $\mathcal{F}$.

Two other points regarding (23) deserve mention. The first is that the Hartree-Fock equation requires solution by self-consistent methods. Since the operators $\mathcal{J}_j$ and $\mathcal{K}_j$ must be calculated for each of the filled orbitals ψ_j, these orbitals are required to obtain $\mathcal{F}$. Thus, to begin a calculation, we must guess a set of ψ_j, use them to calculate the operator $\mathcal{F}$, find its eigenfunctions ψ_i, use the filled ψ_i to calculate a new $\mathcal{F}$ operator, and repeat the whole procedure until self-consistency is achieved. Since the filled ψ_i are used in $\mathcal{F}$ to calculate the average effective electrostatic field in which the electrons in each of the MO's move, the calculation is terminated when the field may be said to have reached self-consistency; hence, the acronym SCF for a calculation employing the Hartree-Fock method.

The second point of interest concerns the eigenvalue ϵ_i, associated with ψ_i. If an electron is removed from ψ_i, it can be shown (Problem 8.1) that the new total energy differs from the old by ϵ_i. Thus, the negative of ϵ_i should be the work necessary to remove an electron from this orbital.[1] This is called Koopmans' theorem and may be stated succinctly as

$$I_i = -\epsilon_i \tag{24}$$

where I_i is the ionization potential of an electron in ψ_i. Thus, ϵ_i may be equated with the energy of an electron in this orbital. We note, however, that if the ϵ_i are summed over all the electrons, the result is

$$\sum_i 2\epsilon_i = \sum_i 2h_i + 2\sum_{ij} 2J_{ij} - K_{ij} \tag{25}$$

which is not the total energy; for on comparison with (5), we see that (25) counts the terms arising from the electron repulsion operator twice. Thus,

---

[1] This assumes that the wavefunctions for all the other electrons remain unchanged, which is obviously an approximation.

the sum of one-electron energies will only approximate the total energy if the nuclear repulsions, which are omitted from (25), are fortuitously the same as the extra electron repulsion terms contained in (25). We will discuss systems in which this is most likely to be the case later in this chapter, when we examine Hückel calculations within the framework of the more rigorous SCF theory.

Problem 8.1.(a)   Use (5) to calculate the energy of the ion formed by the removal of an electron from ψ_i. Show that the energy of the ion differs from that of the neutral molecule by ϵ_i. What is the energy required to remove the second electron from ψ_i? Why is this not equal to ϵ_i?

   (b)   Show that when Ψ_1 differs from Ψ_0 by the excitation of one electron from ψ_i to ψ_j, the fact that $\langle \psi_i | \mathcal{F} | \psi_j \rangle = 0$ implies $\langle \Psi_1 | \mathcal{H} | \Psi_0 \rangle = 0$.

We will now obtain an expression corresponding to (23) for our MO's written as LCAO's and see how to calculate the $\mathcal{F}$ operator. Following Roothan, who first derived the correct formula, we write the MO ψ_i in matrix notation as

$$\langle \psi_i | = [c_{iq}^*] [\phi_q] \quad \text{and} \quad | \psi_i \rangle = [\phi_r] [c_{ri}] \tag{26}$$

where $[\phi_q]$ is a column and $[\phi_r]$ a row matrix. Equations (6)–(9) become

$$h_i = [c_{iq}^*] [\langle \phi_q | \hbar | \phi_r \rangle] [c_{ri}] \equiv [c_{iq}^*] [h_{qr}] [c_{ri}] \tag{27}$$

$$J_{ij} = [c_{iq}^*] [\sum_{s,t} c_{js} c_{jt} (\phi_q \phi_r | \phi_s \phi_t)] [c_{ri}] \equiv [c_{iq}^*] [J_{qr}^j] [c_{ri}] \tag{28}$$

$$K_{ij} = [c_{iq}^*] [\sum_{s,t} c_{js} c_{jt} (\phi_q \phi_t | \phi_s \phi_r)] [c_{ri}] \equiv [c_{iq}^*] [K_{qr}^j] [c_{ri}] \tag{29}$$

$$S_{ij} = [c_{iq}^*] [\langle \phi_q | \phi_r \rangle] [c_{ri}] = [c_{iq}^*] [S_{qr}] [c_{rj}] = \delta_{ij} \tag{30}$$

in which the operators $\hbar$, $\mathcal{J}_j$, $\mathcal{K}_j$, and S are replaced by matrices whose elements are obtained by integrating these operators over the atomic orbitals comprising the MO's. Using the same procedure by which (23) was derived, we find that the total energy is minimized with respect to variations in the c's, and subject to the condition that the MO's remain orthonormal, by an array of c's which satisfies

$$[F_{qr}][c_{ri}] = [S_{qr}][c_{ri}][\epsilon_i] \tag{31}$$

The elements of $[F_{qr}]$ are obtained by evaluating $\mathscr{F}$ over the AO's (i.e., $F_{qr} = \langle\phi_q|\mathscr{F}|\phi_r\rangle$). From (23) and the definitions implicit in (27)–(29)

$$[F_{qr}'] = [h_{qr}] + \sum_j 2[J_{qr}^j] - [K_{qr}^j] \tag{32}$$

Using the definition of bond order for a closed-shell,

$$P_{st} = 2\sum_j c_{js}c_{jt} \tag{33}$$

where the sum again is over just the filled MO's, we can write

$$[F_{qr}] = [h_{qr}] + [\sum_{s,t} P_{st}\{(\phi_q\phi_r|\phi_s\phi_t) - \tfrac{1}{2}(\phi_q\phi_t|\phi_s\phi_r)\}] \tag{34}$$

This is the fundamental equation in LCAO-MO SCF theory; however we can effect a simplification of the above expression for $[F_{qr}]$, since by our assumption of zero differential overlap, many of the integrals in (34) are zero. In fact, the only nonzero integrals are of the form $(\phi_q\phi_r|\phi_s\phi_t)\,\delta_{qr}\,\delta_{st}$ and $(\phi_q\phi_t|\phi_s\phi_r)\,\delta_{qt}\,\delta_{sr}$. Thus, $[F_{qr}]$ reduces to

$$[F_{qr}] = [h_{qr}] + [\sum_s P_{ss}(\phi_q\phi_r|\phi_s^2)\,\delta_{qr} - \tfrac{1}{2}P_{qr}(\phi_q^2|\phi_r^2)] \tag{35}$$

Remembering that $[h_{qr}]$ represents the effect of the one-electron operators (kinetic energy and nuclear-electron attraction) in the Hamiltonian, the diagonal terms of $[F_{qr}]$ are

$$\begin{aligned}
F_{rr} &= h_{rr} + \sum_s P_{ss}(\phi_r^2|\phi_s^2) - \tfrac{1}{2}P_{rr}(\phi_r^2|\phi_r^2) \\
&= -I_r - \sum_{s\neq r}' Z_s'\gamma_{rs} + \tfrac{1}{2}P_{rr}\gamma_{rr} + \sum_{s\neq r} P_{ss}\gamma_{rs} \\
&= -I_r + \tfrac{1}{2}P_{rr}\gamma_{rr} + \sum_{s\neq r}(P_{ss} - Z_s')\gamma_{rs}
\end{aligned} \tag{36}$$

F_{rr} represents the energy of an electron on atom r, which, as (36) shows, depends not only on the ionization potential of r, but also on the number of electrons of opposite spin on r ($\tfrac{1}{2}P_{rr}$) and on the net charges ($P_{ss} - Z_s'$) on all the other atoms.

The off-diagonal elements of $[F_{qr}]$ are

$$F_{qr} = h_{qr} - \tfrac{1}{2}P_{qr}(\phi_q^2|\phi_r^2) = \beta_{qr} - \tfrac{1}{2}P_{qr}\gamma_{qr} \tag{37}$$

We shall discover the physical significance of the second term in (37) in a subsequent section of this chapter.

An actual calculation first requires values of the various integrals, β_{qr} and γ_{qr}, which depend on the molecular geometry. Next an initial set of one-electron wavefunctions is assumed; and by summing the products of coefficients of each pair of AO's over all the MO's that are filled, the bond orders, P_{qr}, and charge densities, P_{rr}, are computed. The matrix $[F_{qr}]$ can then be calculated; and with $[S_{qr}] = [\delta_{qr}]$, (31) is used to find a new set of AO coefficients for each MO. The condition that these coefficients be different from zero is that the determinant

$$|F_{qr} - \epsilon\,\delta_{qr}| = 0 \tag{38}$$

After solving the secular equation, which results from expansion of the determinant, for the orbital energies, the set of AO coefficients corresponding to each energy is found from the ratios of the signed minors of the determinant. If there are N electrons in the system, the $N/2$ MO's with the lowest one-electron energies are chosen for computation of a new set of bond orders and charge densities, from which a new $[F_{qr}]$ can be found. The process of diagonalization of $[F_{qr}]$ and recomputation of its elements is repeated until self-consistency is achieved. At any point in the calculation the total energy, including nuclear repulsions, may be calculated from (5) as

$$
\begin{aligned}
E &= \sum_{q>r,\,r} Z_q' Z_r' \gamma_{qr} + \sum_i [c_{iq}^*]\,[h_{qr} + F_{qr}]\,[c_{ri}] \\
&= \sum_{q>r,\,r} Z_q' Z_r' \gamma_{qr} + \tfrac{1}{2}\sum_{q,\,r} P_{qr}(h_{qr} + F_{qr}) \\
&= \sum_r P_{rr}(-I_r + \tfrac{1}{4}P_{rr}\gamma_{rr}) + \sum_{q>r,\,r} 2P_{qr}\beta_{qr} - \tfrac{1}{2}P_{qr}^2\gamma_{qr} \\
&\quad + (P_{qq} - Z_q')(P_{rr} - Z_r')\gamma_{qr}
\end{aligned}
\tag{39}
$$

which should decrease as the calculation is iterated.

Problem 8.2. Supply the missing steps in the derivation of the last line of (39) from the second. [Hint: Identities you may find useful are $\sum_{q\ne r,\,r} P_{qq} Z_r' = \sum_{q>r,\,r} P_{qq} Z_r' + P_{rr} Z_q'$ and $\sum_{q\ne r,\,r} h_{qr} = \sum_{q>r,\,r} 2h_{qr}$.]

As an illustrative example let us take a simple model for the stabilization of a carbonium ion by an adjacent oxygen. For the oxygen $Z' = 2$, $-I = -33.9$ eV, $\gamma_{oo} = 18.6$ eV; for the carbon $Z' = 1$, $-I = -11.5$ eV, $\gamma_{cc} = 11.1$ eV; and $\beta_{co} = -1.5$ eV, $\gamma_{co} = 8.3$ eV. Let us naively start our calculation with the MO's for ethylene. Since only the bonding MO is filled, $P_{cc} = P_{oo} = P_{co} = 1$. The elements of the $\mathfrak{F}$ matrix are, letting the oxygen be atom 1,

$$F_{11} = -33.9 + \tfrac{1}{2} \times 18.6 = -24.6 \text{ eV} \tag{40}$$

$$F_{22} = -11.5 - 8.3 + \tfrac{1}{2} \times 11.1 = -14.25 \text{ eV} \tag{41}$$

$$F_{12} = F_{21} = -1.5 - \tfrac{1}{2} \times 8.3 = -5.65 \text{ eV} \tag{42}$$

Expanding the determinant

$$\begin{vmatrix} -24.6 - \epsilon & -5.65 \\ -5.65 & -14.25 - \epsilon \end{vmatrix} = \epsilon^2 + 38.85\epsilon + 351 - 32 = 0 \tag{43}$$

which is satisfied by $\epsilon = -27.1$ eV as the lowest root. Substituting this into the determinant in (43) gives the ratio of signed minors

$$\frac{A_{11}}{-A_{12}} = \frac{12.85}{5.65} = 2.28 = \frac{c_1}{c_2} \tag{44}$$

The new wavefunction for the lowest MO is, after normalization,

$$\psi = 0.915\phi_1 + 0.402\phi_2 \tag{45}$$

from which one computes

$$P_{11} = 1.68 \qquad P_{22} = 0.32 \qquad P_{12} = 0.74 \tag{46}$$

The energy of the ground state, calculated from this approximate wavefunction, is

$$E = 1.68(-33.9 + \tfrac{1}{4} \times 1.68 \times 18.6) + 0.32(-11.5 + \tfrac{1}{4} \times 0.32 \times 11.1) \\ - 2 \times 0.74 \times 1.5 - \tfrac{1}{2}(0.74)^2 \times 8.3 + 0.32 \times 0.68 \times 8.3 = -49.9 \text{ eV} \tag{47}$$

compared with

$$E = -33.9 + \tfrac{1}{4} \times 18.6 - 11.5 + \tfrac{1}{4} \times 11.1 - 2 \times 1.5 - \tfrac{1}{2} \times 8.3$$
$$= -45.1 \text{ eV} \tag{48}$$

for the starting ethylenic wavefunction. Clearly, the much greater electronegativity of the oxygen tends to keep the two electrons localized on this atom so that our initial choice of an ethylenic wavefunction, which partitions the electrons equally between the two atoms, was not very realistic.

Problem 8.3. Show that the approximate wavefunction which keeps both electrons completely localized on the oxygen atom is only slightly higher in energy than the wavefunction (45), obtained after one SCF iteration when an ethylenic wavefunction is used to compute the intial $\mathfrak{F}$ matrix.

Problem 8.4. Carry out another SCF iteration, employing (45) to calculate the $\mathfrak{F}$ matrix. Use the derived wavefunction to calculate the net stabilization energy of the carbonium ion by the adjacent oxygen ion pair. Show that despite this substantial stabilization, very little π charge is calculated to be transferred from oxygen to carbon.

It is clear from the amount of work necessary to do an SCF calculation on a π system consisting of two atomic orbitals that for a molecule of any size an SCF calculation requires the use of a computer. Writing a program to carry out such a calculation is an excellent exercise both in programming and in understanding the mechanics of carrying out an SCF calculation. The input consists of a set of effective nuclear charges, Z'_r, ionization potentials, I_r, β_{qr} and γ_{qr} integrals over the AO's, and a set of starting orbitals. The program calculates the $\mathfrak{F}$ matrix, diagonalizes it, finds a new set of MO's, and computes the total energy, iterating the procedure until self-consistency is achieved. The output consists of the self-consistent MO's, orbital energies, bond orders, charge densities, and the total energy. Usually, computing centers have standard programs for diagonalizing matrices—finding the eigenvalues (MO energies) and associated eigenvectors (AO coefficients for each MO)—and once the main program has set up the $\mathfrak{F}$ matrix, the diagonalization program may be called as a subroutine to operate on it. Most of the running time of an

LCAO-MO-SCF program that employs the semiempirical Pariser-Parr-Pople methodology is spent in matrix diagonalization.

Symmetry in SCF Theory

Even if one has access to a very large digital computer, the time spent diagonalizing the $n \times n$ matrix for a calculation involving n AO's can be considerable. Since the time it takes to diagonalize a matrix by computer is roughly proportional to n^2 or n^3, it may be advantageous to try to reduce the problem of diagonalization of an $n \times n$ matrix to the diagonalization of several smaller matrices. As an example, if we wish to find the SCF π MO's for butadiene we must diagonalize a 4×4 matrix. If we do the problem by hand we must expand the determinant, which gives $4! = 24$ terms, and then solve a quartic equation. However, the problem can be reduced through the use of symmetry to the diagonalization of two 2×2 matrices. If we use the two sets of basis functions $(\phi_1 + \phi_4)$, $(\phi_2 + \phi_3)$ and $(\phi_1 - \phi_4)$, $(\phi_2 - \phi_3)$, because they belong to different representations of the molecular symmetry group, there are no matrix elements of $\mathfrak{F}$ between sets. Thus, using these symmetry orbitals instead of AO's as a basis set, the problem is reduced to

$$\begin{bmatrix} F'_{11} & F'_{12} & 0 & 0 \\ F'_{21} & F'_{22} & 0 & 0 \\ 0 & 0 & F'_{33} & F'_{34} \\ 0 & 0 & F'_{43} & F'_{44} \end{bmatrix} \begin{bmatrix} c_1 \\ c_2 \\ c_3 \\ c_4 \end{bmatrix} = \begin{bmatrix} \epsilon & 0 & 0 & 0 \\ 0 & \epsilon & 0 & 0 \\ 0 & 0 & \epsilon & 0 \\ 0 & 0 & 0 & \epsilon \end{bmatrix} \begin{bmatrix} c_1 \\ c_2 \\ c_3 \\ c_4 \end{bmatrix} \quad (49)$$

where the primed matrix elements are evaluated over the symmetry orbitals. Solution of (49) requires only diagonalizing the two 2×2 submatrices.

Let us actually set up and diagonalize the two matrices for the symmetry orbitals in the first iteration of an SCF calculation on *transoid* butadiene, starting with the Hückel orbitals. For the MO which is symmetric about the C_2 symmetry axis we seek a solution of

$$\mathfrak{F}[c_1(\phi_1 + \phi_4) + c_2(\phi_2 + \phi_3)] = \epsilon[c_1(\phi_1 + \phi_4) + c_2(\phi_2 + \phi_3)] \quad (50)$$

The $\mathfrak{F}$ matrix elements are just the coefficients of the c's in the two equations that we obtain after multiplying through by ϕ_1 and ϕ_2 and integrating.

To calculate the matrix elements from (36) and (37) we need the charge densities and bond orders evaluated from our assumed starting Hückel wavefunction.

All the charge densities are, of course, one in Hückel theory, since butadiene is an AH. The bond orders calculated in Chapter 3, were $P_{12} = 0.89$, $P_{13} = 0$, $P_{14} = -0.45$, and $P_{23} = 0.45$. The vanishing of the bond order between carbon atoms belonging to the same set (starred or unstarred) is, of course, also a property of neutral AH's. Since the ionization potential of a carbon atom occurs in every term on the diagonal of the $\mathfrak{F}$ matrix, we can omit it from the calculation by defining all our π electron energies relative to that of an electron in an isolated valence state carbon $2p$ orbital. Taking

$$\beta_{12} = -1.8 \text{ eV} \quad \text{and} \quad \beta_{23} = -1.4 \text{ eV},$$

$$\gamma_{11} = 11.1 \text{ eV}, \ \gamma_{12} = 7.7 \text{ eV}, \ \gamma_{23} = 7.4 \text{ eV} \quad \text{and} \quad \gamma_{14} = 4.0 \text{ eV}$$

we compute the elements, F'_{qr}, of the matrix for the symmetry orbital basis set in terms of the elements F_{qr}, evaluated over the AO's as follows:

$$F'_{11} = \langle \phi_1 | \mathfrak{F} | \phi_1 + \phi_4 \rangle = F_{11} + F_{14}$$
$$= \tfrac{1}{2} \times 11.1 + \tfrac{1}{2} \times 0.45 \times 4.0 = 6.5 \text{ eV} \tag{51}$$

$$F'_{22} = \langle \phi_2 | \mathfrak{F} | \phi_2 + \phi_3 \rangle = F_{22} + F_{23}$$
$$= \tfrac{1}{2} \times 11.1 - 1.4 - \tfrac{1}{2} \times 0.45 \times 7.4 = 2.5 \text{ eV} \tag{52}$$

$$F'_{12} = \langle \phi_1 | \mathfrak{F} | \phi_2 + \phi_3 \rangle = F'_{21} = F_{12} + F_{13} = F_{12}$$
$$= -1.8 - \tfrac{1}{2} \times 0.89 \times 7.7 = -5.2 \text{ eV} \tag{53}$$

The resulting matrix equation is

$$\begin{bmatrix} 6.5 & -5.2 \\ -5.2 & 2.5 \end{bmatrix} \begin{bmatrix} c_1 \\ c_2 \end{bmatrix} = \begin{bmatrix} \epsilon & 0 \\ 0 & \epsilon \end{bmatrix} \begin{bmatrix} c_1 \\ c_2 \end{bmatrix} \tag{54}$$

Carrying out the diagonalization, the lowest MO is found to be $\psi_1 = 0.40(\phi_1 + \phi_4) + 0.58(\phi_2 + \phi_3)$ with $\epsilon = -1.1$ eV relative to an electron in an isolated carbon $2p$ orbital.

Problem 8.5.   Find the other filled SCF-MO starting with symmetry orbitals. Compute the total energy of butadiene, relative to four isolated carbon atoms. Show, as outlined below, that in a neutral closed-shell AH like butadiene, the consequences of the pairing theorem still hold for the SCF-MO's: Demonstrate that in butadiene the SCF-MO's come in pairs with equal and opposite orbital energy, relative to $F_{rr} = -I_r + \frac{1}{2}\gamma_{rr}$, and with coefficients at each atom equal in magnitude but opposite in sign at all unstarred atoms. Show that the π charge density on each atom is one and the bond orders between atoms of the same set vanish.

Problem 8.6.   Set up and carry out an SCF calculation on the allyl cation. Unlike the Hückel result, you should find that the central carbon bears a small *negative* charge. What term in the expression for F_{22} is responsible for the appearance of a negative charge here? How do the elements of the initial $\mathfrak{F}$ matrix change in going from the allyl cation to the anion? Without carrying out an SCF calculation on the anion, predict the distribution of charge found in it. Explain qualitatively the physical reason why charge alternation is favored in these ions.

Problem 8.7.   Use a computer to assist you in demonstrating that Hückel theory overestimates the dipole moment in pentatriafulvalene (Chap. 3, p. 74). Write a program to carry out an HMO calculation on this molecule with all β assumed equal. Use (11) and (12) of Chapter 3 to make your Hückel calculation self-consistent with respect to the molecular geometry. Use the final geometry to compute the required integrals for an SCF calculation, and begin the computation using the self-consistent Hückel HMO's. In your SCF calculation, for simplicity, you need not adjust the γ's but only the β's for changes in molecular geometry. Compare the three geometries and dipole moments predicted by your calculations.

In the event that you have access to only a relatively slow computer, you may find it worthwhile to take advantage of the symmetry of pentatriafulvalene to reduce the diagonalization from that of an 8×8 to that of a 3×3 and a 5×5 matrix. Most diagonalization subroutines will handle only Hermitian or symmetric matrices. Show that to obtain such a matrix, a normalized symmetry basis set, $\phi_i' = (1/\sqrt{2})(\phi_i + \phi_j)$, must be used, where ϕ_i and ϕ_j are symmetry equivalent orbitals.

Comparison of Hückel and SCF Theories

Although in a hydrocarbon in which all the MO's are determined by symmetry, Hückel and SCF theory must give the same MO's, it may seem surprising that in butadiene, where the MO's are not completely symmetry-determined, the SCF MO's are virtually identical to those obtained by the Hückel calculation in Problem 3.6. Let us see why this is the case by comparing the elements in the Hückel $\mathcal{H}$ and the SCF $\mathcal{F}$ matrix, which represent the one-electron operators whose eigenfunctions are the MO's. Taking integrals over AO's, the diagonal elements of the Hückel $\mathcal{H}$ matrix are all α. In SCF theory they are given by (36) as

$$F_{rr} = -I + \tfrac{1}{2}P_{rr}\gamma_{rr} + \sum_{q \neq r}(P_{qq} - Z'_q)\gamma_{qr}$$

But in butadiene or any other closed-shell neutral alternant $P_{qq} - Z'_q$ is zero for each atom, since all the atoms in the molecule are neutral. Thus, in such a molecule

$$F_{rr} = -I + \tfrac{1}{2}\gamma_{rr} \qquad (55)$$

which is just what we found α to be in Chapter 1. Moreover, so long as all the atoms bear no net charge, all the diagonal elements in the $\mathcal{F}$ matrix are identical and play no part in determining the MO's, as is the case for *all* hydrocarbons in Hückel theory. The off-diagonal elements in the $\mathcal{H}$ matrix are β for atoms that are bonded and zero for those that are not. In Chapter 3 we found that around the C=C bond distance, β is proportional to the bond length, which is in turn proportional to the bond order. In fact we found in Equation (24) of Chapter 3

$$\beta_{qr}^H = \beta_0^H - 0.48(1 - P_{qr})\beta_0^H$$

In SCF theory the off-diagonal elements are given by (37) as

$$F_{qr} = \beta_{qr} - \tfrac{1}{2}P_{qr}\gamma_{qr}$$

For atoms bonded together γ_{qr} will be constant to within about 10%, so the off-diagonal elements for such atoms will have the same type of dependence on the bond order as in Hückel theory. However, unlike the Hückel $\mathcal{H}$ matrix, in the SCF $\mathcal{F}$ matrix the off-diagonal elements between nonbonded atoms are not zero but are equal to $-\tfrac{1}{2}P_{qr}\gamma_{qr}$. Nevertheless, for a certain class of molecules, namely the alternant hydrocarbons, this may not result in a very big difference between Hückel and SCF-MO's. The reason is that in a neutral closed-shell AH, the bond orders between

AO's of the same set vanish. This means that in these molecules the only interactions in the $\mathfrak{F}$ matrix, not present in the Hückel $\mathfrak{IC}$ matrix, are between atoms separated by two or more other atoms. Consequently, these matrix elements, which depend on γ_{qr}, are small compared to those for bonded atoms. For instance, in the calculation on *transoid* butadiene, F_{14}, the element neglected in the $\mathfrak{IC}$ matrix, is 0.9 eV. This repulsion tends to move electron density in the lowest MO off the terminal carbons; but the effect is so small that if this matrix element is omitted in the foregoing calculation, the lowest MO is $\psi = 0.42(\phi_1 + \phi_4) + 0.57(\phi_2 + \phi_3)$, which differs very little from the MO obtained when F_{14} is included.

We see then that for closed-shell neutral AH's the SCF and Hückel methods can be expected to give very similar sets of MO's; but what about total energies? We remarked in connection with (25) that the Hückel theory expression for the total energy as a sum of orbital energies was incorrect, because it counted the electron repulsion terms twice. This will only work if the nuclear repulsion terms, which Hückel theory also ignores, are of the same magnitude as the extra electron repulsion terms which Hückel theory includes. In a neutral AH, because the individual atoms are also neutral, the repulsions between the nuclei are expected to be roughly the same as those between the electrons. Therefore, one might expect Hückel theory to have the most success with total energies in this class of molecules. However, even here the cancellation will not be perfect, because included in the terms that arise from the two-electron operator are repulsions between electrons in the same AO and corrections, arising from the exchange integrals, to the repulsion energy calculated from the overall charge distribution. We can see this more clearly in the correct expression for the total energy. For a neutral AH the electrostatic terms in (39) vanish and the equation becomes

$$E = \sum_r - I_r + \tfrac{1}{4}\gamma_{rr} + \sum_{q>r,r} 2P_{qr}(\beta_{qr} - \tfrac{1}{4}P_{qr}\gamma_{qr}) \qquad (56)$$

The corresponding expression in Hückel theory is from Equation (25) of Chapter 3,

$$E_H = \sum_r \alpha_r + \sum_{q>r,r} 2P_{qr}\beta_{qr}^H$$

If we set $\alpha_r = -I_r + \tfrac{1}{2}\gamma_{rr}$, comparison of the two equations with $\beta_{qr}^H = \beta_{qr}$ shows they differ by

$$E_H - E = \tfrac{1}{4}\sum \gamma_{rr} + \tfrac{1}{2}\sum_{q>r,r} P_{qr}^2\gamma_{qr} \qquad (57)$$

Hückel theory giving a considerably greater energy than the true one. In fact, if we are interested in calculating total energies in Hückel theory, we do better with $\alpha = -I$. Since in a hydrocarbon the diagonal elements of the $\mathcal{K}$ matrix do not play a role in determining the MO's, we are free to choose α to best fit total energies. With $\alpha = -I$ the difference between the Hückel theory expression and the correct SCF one is

$$E_H - E = -\tfrac{1}{4} \sum_r \gamma_{rr} + \tfrac{1}{2} \sum_{q>r,r} P^2_{qr}\gamma_{qr} \tag{58}$$

which allows for some cancellation of the two errors, mentioned above, that HMO theory makes in tacitly equating nuclear and electronic repulsions. In general, the first term is larger than the second, so with $\alpha = -I$ Hückel theory overestimates the stability of the π system relative to isolated atoms. For instance, for butadiene $E_H - E$ in (58) is -3.8 eV; so Hückel theory overestimates the π bond energy by nearly 90 kcal/mole with $\alpha = -I$ and $\beta^H_{qr} = \beta_{qr}$. However, the Hückel and SCF β's need not be the same; and if β^H is chosen empirically, judicious parameter selection can give better results. Note that for

$$\alpha = -I_r + \tfrac{1}{4}\gamma_{rr} \quad \text{and} \quad \beta^H_{qr} = \beta_{qr} - \tfrac{1}{4}P_{qr}\gamma_{qr}$$

in Hückel theory, (25) of Chapter 3 becomes identical with the energy of a neutral AH given by (56), except for the nonzero terms in the latter equation for atoms not bonded to each other and belonging to different sets. The sum of such terms can, nevertheless, be quite large; and it is hard to see, purely theoretically, what choice of α and β will, in general, bring (25) of Chapter 3 into exact agreement with (56). Of course, we really have no reason for expecting that this should be possible, for we found that Hückel theory has quantitative success in calculating *energy differences* between AH's, rather than total energies. This success is probably due to nearly identical discrepancies in two molecules between the Hückel and SCF expressions for total energies. When the difference in energy between the molecules is computed, the discrepancies tend to cancel; and, presumably any residue is taken care of systematically in the semiempirical value chosen for β^H.

Problem 8.8. Use (58) to calculate the difference between the Hückel and SCF expressions for the total energy of benzene with $\alpha = -I$ and $\beta^H = \beta$. Do the same for three ethylene molecules. Compute the difference between the Hückel and SCF expressions for the benzene resonance energy. Use Equation (30) of Chapter 7 to compute the repulsion integrals and confirm the statement in Chapter 3 that the electron repulsion in benzene is about 16 kcal/mole greater than that in the fictitious cyclo-hexatriene.

Electron Correlation

In an LCAO approximation the correlation of the motions of two π electrons may be roughly divided into two types—that within AO's, known as vertical correlation, and correlation of occupancy of AO's, termed horizontal. By choosing semiempirical values for our electron repulsion integrals, we are allowing for correlation of both kinds; since even SCF theory does not correlate the motions of electrons with different spins so that they simultaneously tend to occupy different AO's. All SCF theory can do for electrons of opposite spin is give an optimum *average* charge distribution via the dependence of the diagonal terms in the $\mathfrak{F}$ matrix on not only the ionization potential of each atom, but also on the electron density already on the atom and the net charges distributed over the rest of the molecule. This should not be confused with electron correlation, which SCF theory provides to some extent for electrons of the same spin via the term $-\frac{1}{2}P_{qr}\gamma_{qr}$ in the off-diagonal elements of the $\mathfrak{F}$ matrix. If one traces the origin of these terms, one finds that they arise from the exchange operator in $\mathfrak{F}$. This operator corrects the coulombic repulsion energy between two electrons for the fact that if they have the same spin, they tend to avoid each other.[2] Therefore, bond orders must not only be associated with bond formation between two atoms but also with correlation of the motions of electrons of the same spin.

Let us see why this is so. Suppose we have a many-electron wave-function $\Psi = |\psi_1^\alpha \psi_1^\beta \ldots \psi_n^\alpha \psi_n^\beta\rangle$ and we wish to find the probability of an electron being at atom q. We can define a one-electron operator $\delta(q)$ which has the effect

$$\delta(q)\phi_q = \phi_q \qquad \delta(q)\phi_r = 0 \quad r \neq q \qquad (59)$$

and so picks out ϕ_q from an MO. The probability of finding an electron at atom q can be obtained from a sum of such operators, one for each electron. Since there are two electrons in each filled MO, we obtain

---

[2] For electrons in the same orbital, ψ_i, the exchange operator also corrects the electron repulsion, calculated from the Coulomb operator, from $2J_{ii}$ to $2J_{ii} - K_{ii} = J_{ii}$.

$$\langle \Psi \mid \sum_{\substack{i \\ \text{electrons}}} \delta(q) \mid \Psi \rangle = 2 \sum_{\substack{i \\ \text{filled MO's}}} \langle \psi_i \mid \delta(q) \mid \psi_i \rangle = 2 \sum_{\substack{i \\ \text{filled MO's}}} \langle \psi_i \mid c_{iq}\phi_q \rangle = 2 \sum_{\substack{i \\ \text{filled MO's}}} c_{iq}^2 = P_{qq}$$

$$(60)$$

which is, of course just the number of electrons at atom q. Suppose now that we wish to know the probability of finding an electron of α spin at q. We define $\delta^\alpha(q)$, which operates just like $\delta(q)$, but gives a zero unless the electron has spin α. We obtain

$$\langle \Psi \mid \sum_{\substack{i \\ \text{electrons}}} \delta^\alpha(q) \mid \Psi \rangle = \sum_i \langle \psi_i^\alpha \mid \delta^\alpha(q) \mid \psi_i^\alpha \rangle + \langle \psi_i^\beta \mid \delta^\alpha(q) \mid \psi_i^\beta \rangle = \sum_{\substack{i \\ \text{filled MO's}}} c_{iq}^2 = \tfrac{1}{2} P_{qq}$$

$$(61)$$

since only half the electrons have α spin. We can analogously define a product of sums of one-electron operators $\sum \delta(q) \sum \delta(r)$ which tell us the probability of finding an electron at q and another simultaneously at r. For finding electrons of opposite spin, α at q and β at r, the probability is

$$\langle \Psi \mid \sum_{\substack{\text{electrons}}} \delta^\alpha(q) \sum \delta^\beta(r) \mid \Psi \rangle = \langle \dots \psi_i^\alpha \dots \psi_j^\beta \mid \sum_{\substack{\text{electrons}}} \delta^\alpha(q) \sum \delta^\beta(r) \mid \dots \psi_i^\alpha \dots \psi_j^\beta \rangle$$

$$= (\sum_{\substack{i \\ \text{filled MO's}}} \langle \psi_i^\alpha \mid \delta^\alpha(q) \mid \psi_i^\alpha \rangle)(\sum_{\substack{j \\ \text{filled MO's}}} \langle \psi_j^\beta \mid \delta^\beta(r) \mid \psi_j^\beta \rangle)$$

$$= \tfrac{1}{4} P_{qq} P_{rr} \qquad (62)$$

which is of course just the product of the separate probabilities. Therefore, the probability of an electron of α spin being at q is unaffected by there being one of β spin at r. However, for two electrons of α spin, noting that they cannot be in the same MO, we find

$$\langle \Psi \mid \sum_{\substack{\text{electrons}}} \delta^\alpha(q) \sum \delta^\alpha(r) \mid \Psi \rangle = \langle \dots \psi_i^\alpha \dots \psi_j^\alpha \mid \sum_{\substack{\text{electrons}}} \delta^\alpha(q) \sum \delta^\alpha(r) \mid \dots \psi_i^\alpha \dots \psi_j^\alpha \rangle$$

$$= \sum_{\substack{i \neq j,\, j \\ \text{filled MO's}}} \langle \psi_i^\alpha \mid \delta^\alpha(q) \mid \psi_i^\alpha \rangle \langle \psi_j^\alpha \mid \delta^\alpha(r) \mid \psi_j^\alpha \rangle - \sum_{\substack{i \neq j,\, j \\ \text{filled MO's}}} \langle \psi_i^\alpha \mid \delta^\alpha(q) \mid \psi_j^\alpha \rangle \langle \psi_j^\alpha \mid \delta^\alpha(r) \mid \psi_i^\alpha \rangle$$

$$(63)$$

$$= \sum_{i \neq j,\, j} c_{iq}^2 c_{jr}^2 - \sum_{i \neq j,\, j} c_{iq} c_{jq} c_{jr} c_{ir} = \sum_{i,\, j} c_{iq}^2 c_{jr}^2 - \sum_{i,\, j} c_{iq} c_{jq} c_{jr} c_{ir}$$

$$= \sum_{\substack{i \\ \text{filled MO's}}} c_{iq}^2 \sum_j c_{jr}^2 - \sum_{\substack{i \\ \text{filled MO's}}} c_{iq} c_{ir} \sum_j c_{jq} c_{jr} = \tfrac{1}{4}(P_{qq} P_{rr} - P_{qr}^2)$$

Note that the change in the summation in (63) from $i \neq j$ to all combinations of i, j is made possible by the fact that $c_{iq}^2 c_{jr}^2 = c_{iq}c_{jq}c_{jr}c_{ir}$ for $i = j$. Thus, in (63) $P_{qq}P_{rr}$ contains a term that corresponds to having two electrons of the same spin in the same orbital, which is cancelled by the presence of the same term in P_{qr}^2. This is the same trick that allows us to write the repulsion between two electrons in the same orbital as $J_{ii} = 2J_{ii} - K_{ii}$.

We see that (63) differs from the product of the separate probabilities by $-\frac{1}{4}P_{qr}^2$. Thus, there is a low probability of simultaneously finding two electrons of the same spin on atoms between which there is a large bond order. It is relatively unlikely, therefore, that two electrons of the same spin will simultaneously be on atoms that are bound to each other, and the probability that they will simultaneously be on the same atom is, of course, zero. For alternant hydrocarbons, since $P_{qr} = 0$ for atoms of the same set, it is most probable to find electrons of the same spin on atoms of the same set. For instance, in butadiene or any other neutral AH, because all the π charges are unity, the probability of finding an electron of α spin at any atom q is $\frac{1}{2}$. The probability of simultaneously finding another α spin electron at atom r is $\frac{1}{2}(1 - P_{qr}^2)$. Using the SCF wavefunction for butadiene we can calculate that if there is a π electron of α spin at atom 1, then the probability of finding the other π electron of α spin at atom 2 is 0.07, at atom 3, 0.50, and at atom 4, 0.43.

Note that the computation of the total energy takes account of this correlation of the motions of electrons of the same spin in an antisymmetrized wavefunction through the term $-\frac{1}{2}P_{qr}^2\gamma_{qr}$. This is just the correction, arising from the exchange integrals K_{ij}, to the electron repulsion energy for electrons of the same spin. The Coulomb repulsion energy, calculated from the J_{ij} integrals, assumes that the probability of finding any two electrons simultaneously at q and r is $P_{qq}P_{rr}$. For electrons of the same spin, the exchange integrals correct this by twice the correlation function, $-\frac{1}{4}P_{qr}^2$, once for the set of electrons with α spin and once for those with β spin. Thus, the use of a correctly antisymmetrized many-electron wavefunction automatically provides for correlation between electrons of the same spin. Moreover, since the exchange operator appears in $\mathfrak{F}$, the set of MO's one obtains by an SCF calculation will tend to optimize correlation of these electrons. However, as we have seen, no wavefunction constructed from a single Slater determinant, even though the determinantal wavefunction be obtained from an SCF calculation, allows for correlation between electrons of opposite spins. We have seen in the last chapter that such electron correlation can be introduced through configuration interaction by writing the wavefunction as a sum of Slater determinants. In the next chapter we shall see how such a sum can be obtained within the SCF formalism.

Problem 8.9.   Despite the fact that Hückel theory predicts the same first ionization potential for both ethylene and benzene, that of the former is found experimentally to be about 1.3 eV larger than the latter. Use the equation $\epsilon = \langle \psi | \mathcal{F} | \psi \rangle$ to calculate the negative of the ionization potentials for both these molecules. Explain the origin of most of the difference between the orbital energies of the highest occupied MO of each. Allowing for the difference in β in these two molecules, how well does SCF theory do in quantitatively accounting for the observed difference in ionization potentials? [Hint: It is easiest to use the benzene orbital that has a node through two atoms and consequently resembles the antisymmetric combination of two bonding ethylene π MO's.]

Problem 8.10.(a)   Suppose that there is an α spin electron in one of the p orbitals in ethylene. What is the probability distribution for the remaining π electron? Account for the term $\frac{1}{2}(\gamma_{11} - \gamma_{12})$ in the orbital energy of ethylene as resulting from the net repulsions and attractions experienced by the α spin electron.

(**b**)   Suppose there is an α spin π electron at one atom in benzene. Compute the probability distributions for the other π electrons of α and β spin. Show that the energy arising from the net repulsions and attractions to which the α spin electron is subjected is equal to the average of the electrostatic terms in the orbital energies of ψ_1, ψ_2, and ψ_3 in benzene.

(**c**)   From (5) and (23) the electronic energy of a closed-shell molecule can be written as $\sum \epsilon_i + h_i$, where the sum is over the filled orbitals. Use this fact and the results in parts (a) and (b) to account for the electrostatic terms in the expression for the π energy difference between benzene and three ethylene molecules. Write resonance structures corresponding to the terms in the benzene wavefunction where two electrons are found in the same AO, and show that the Coulomb energy is higher in benzene than in three ethylenes because charge is separated farther in the ionic terms of the wavefunction for the former molecule.

Summary

In this chapter we derived the SCF equations for finding as LCAO-MO's, the set of one-electron wavefunctions that minimize the energy of a closed-shell ground state. It proved expedient to represent the $\mathcal{F}$ operator

248 Molecular Orbitals from Self-Consistent Field Theory

by the matrix $[F]$, whose elements are found by evaluating $\mathscr{F}$ over the AO basis set. The MO's are found by diagonalizing the $\mathscr{F}$ matrix, and its eigenvalues are the orbital energies. We saw that the time spent diagonalizing the matrix can be reduced by using symmetry orbitals as a basis set, since orbitals which form the basis for different representations of the molecular symmetry group will have no matrix element of $\mathscr{F}$ between them.

The fact that both the Hückel $\mathscr{H}$ and the SCF $\mathscr{F}$ are one-electron operators led us to explore the possibility that the success of the former theory could be attributed to the fact that, for certain systems, the SCF matrix, $[F]$, reduces to $[H]$. Indeed, for closed-shell neutral AH's—just those molecules where Hückel theory works best—the form of the two is identical, except for some small off-diagonal terms in the $\mathscr{F}$ matrix which are omitted in the $\mathscr{H}$ matrix. Therefore, the MO's for neutral AH's, obtained from Hückel and SCF theory, should be very similar. Moreover, in the calculation of total energies for this class of molecules, a correct semiempirical choice for the Hückel β can very roughly approximate the results obtained by SCF theory. This only comes about through a fortuitous partial cancellation of electron and nuclear repulsion energies. By summing the orbital energies twice to get the total energy, Hückel theory counts the electron repulsions twice, but this is more or less balanced by its neglect of the nuclear repulsions. Even so, we saw that the balancing of these two errors, implicit in HMO theory, is not complete. Consequently, Hückel theory should be best at predicting energy differences between molecules, so that cancellation of the errors in the energy of each can occur. It is to be emphasized that only for neutral closed-shell AH's, where each atom is neutral, can Hückel theory be expected to give reasonable MO's or even approximate total energy differences. For polar molecules the cancellations, which for neutral AH's result in a close resemblance between the SCF $\mathscr{F}$ matrix and the $\mathscr{H}$ matrix, as well as a similarity between the correct expression for the total energy and the Hückel expression, do not occur. Especially in ions, where the total number of protons and electrons is different, the electron repulsion certainly will not be nearly equal to the repulsion between the effective nuclear charges; and Hückel theory, particularly in the calculation of total energies, is not to be trusted.

Problem 8.11. Olah has prepared the dications of some derivatives of cyclobutadiene. Predict, using Hückel theory, whether the dication of trimethylenemethane should be more or less stable than that of cyclobutadiene. For simplicity, assume that the bond lengths in both molecules are the same. Make the same prediction regarding the relative stabilities of these ions, using (39) to calculate the energy of each. Take all bond lengths as 1.43 Å and use $\beta = -1.5$ eV.

In our discussion of SCF wavefunctions we remarked that, although such a wavefunction gives an optimum charge distribution and also augments the correlation between electrons of the same spin, it provides no correlation between electrons of opposite spin. We also noted in our discussion of correlation between electrons of the same spin in an antisymmetrized wavefunction that not only is the bond order a measure of the amount of bonding between two atoms, if they are nearest neighbors; but, in general, the magnitude of the bond order between any two atoms is an index of the amount of correlation between electrons of the same spin on them.

FURTHER READING

LIONEL SALEM, *The Molecular Orbital Theory of Conjugated Systems*, Benjamin, New York, 1966. Chapter 2 discusses SCF calculations on hydrocarbon ions and the pairing theorems for AH ions.

The reader now has sufficient background to read a recent review article on localized orbitals by W. ENGLAND, L. S. SALMON, and K. RUE-DENBERG, *Fort. Chem. Forsch.* **23**, 31 (1971).

9

Wavefunctions for Open-Shell Pi Systems

In Chapter 8 we saw that an SCF wavefunction allows for some correlation of the motions of electrons of the same spin but makes no provision for correlation between electrons of opposite spin. Such correlation can be introduced, as we have already seen, by carrying out configuration interaction. It can, however, also be obtained completely within the SCF formalism by constructing different sets of SCF orbitals for electrons of opposite spin.

The physical reason for expecting electrons of opposite spin to have different wavefunctions is as follows. Electrons of the same spin already have well-correlated motions through antisymmetrization of the many-electron wavefunction. Thus, even though two MO's for electrons of the same spin have large coefficients at the same set of atoms, the electrons tend to avoid each other. However, electrons of opposite spin do not have their motions correlated by antisymmetrization; therefore, the repulsion between two such electrons will be minimized only if their MO's do not both have large coefficients at the same atom. In Chapter 5 we discussed experimental evidence which suggests that electrons of opposite spin do indeed have different wavefunctions, for the spin densities in the allyl radical can most easily be interpreted in terms of different wavefunctions for the two electrons which occupy ψ_1 in the Hückel description of the radical.

Although the Different Orbitals for Different Spins (DODS) SCF approach can be used in closed-shell systems, its most important application is to molecules containing an open-shell. The reason is that in a closed-shell system, although the usual SCF calculation will fail to provide for adequate correlation between electrons of opposite spin, this deficiency can be remedied in a semiempirical calculation by judicious parameter selection. In fact, Dewar has shown[1] that semiempirical SCF calculations give heats of formation for closed-shell pi systems that are in excellent agreement with experiment. Since, as we shall see, DODS-SCF calculations are more complicated to carry out and the resulting wavefunctions more difficult to interpret, the utilization of this method for semiempirical calculations on closed-shell systems is not only unnecessary but probably also undesirable.

Dewar has also shown that a closed-shell SCF approach to calculations on open-shell pi systems yields heats of formation of conjugated radicals that agree well with experiment, and the difference between open- and closed-shell SCF equations provides a good starting point for our discussion of radicals.

Dewar's Half-Electron Method

Dewar's approach to treating open-shell systems by a closed-shell formalism, for the purpose of calculating total energies, utilizes the similarity between the SCF equations for a radical and those for a fictitious closed-shell system in which the odd electron is replaced by two half-electrons. For example, in a radical in which allowance for DODS is not made the total energy is given by

$$E = 2 \sum_i h_i + h_{i'} + \sum_{i,j} 2J_{ij} - K_{ij} + \sum_j 2J_{i'j} - K_{i'j} \qquad (1)$$

[1] See M.J.S. Dewar, *The Molecular Orbital Theory of Organic Chemistry*, McGraw-Hill, New York, 1969, Chapter 5.

where the sums are over the doubly filled orbitals, and the prime denotes the singly occupied orbital. When this orbital is occupied by two half-electrons of opposite spin, an identical expression is obtained, except for the addition of $\frac{1}{4}J_{i'i'}$ for the repulsion between the two half-electrons.

The problem in dealing with a radical in SCF theory is that the Fock equations for finding the MO's are different for electrons of opposite spin. For instance, if the unpaired electron has α spin, then the orbital energies for the two electrons in ψ_i with spin α and β are given respectively by

$$\epsilon_i^\alpha = h_i + \sum_{j \neq i'} 2J_{ij} - K_{ij} + J_{ii'} - K_{ii'} \tag{2}$$

$$\epsilon_i^\beta = h_i + \sum_{j \neq i'} 2J_{ij} - K_{ij} + J_{ii'} \tag{3}$$

From these equations it is clear that the electrons in ψ_i move in different potentials depending on their spin; therefore, even an SCF calculation in which both electrons are initially placed in the same ψ_i must necessarily lead to different wavefunctions for the electrons. Thus, an SCF calculation on a radical *requires* that we allow electrons of different spins to occupy different orbitals, and we will explore the consequences of this fact shortly.

In contrast, the expression for the orbital energy in the fictitious system containing two half-electrons is the same for α or β spin and, except for the half-electrons, is

$$\epsilon_i = h_i + \sum_{j \neq i'} 2J_{ij} - K_{ij} + J_{ii'} - \tfrac{1}{2}K_{ii'} \tag{4}$$

The orbital energy of the half-electrons is just half of (4), namely

$$\epsilon_{i'} = \tfrac{1}{2}h_{i'} + \tfrac{1}{2}\Big(\sum_{j \neq i'} 2J_{i'j} - K_{i'j}\Big) + \tfrac{1}{2}J_{i'i'} - \tfrac{1}{4}K_{i'i'} \tag{5}$$

Equation (4) is just the average of (2) and (3). Because in the radical we have assumed that the unpaired electron in $\psi_{i'}$ has spin α, ϵ_i^β is zero in (3) for $\psi_{i'}$. Therefore, (5) is also the average of (2) and (3) except for the term, $\tfrac{1}{2}J_{i'i'} - \tfrac{1}{4}K_{i'i'} = \tfrac{1}{4}J_{i'i'}$ for the repulsion between the fictitious half-electrons. Therefore, with the exception of this term in the potential for the half-electrons in $\psi_{i'}$, the electrons in the model move in a potential which is the average of that experienced by the electrons in the radical. Thus, the *overall* electron distribution calculated for the half-electron model should be very similar to that computed for the radical. Of course, by averaging the potentials in which electrons of opposite spin move, we lose the correlation effects that give rise to negative spin density. Nevertheless, implicitly including these effects in our choice of parameters should work as well for calculating the total energies of radicals from the half-electron model as it does in computing the total energies of closed-shell systems. It is understandable, therefore, that Dewar, employing the same set of parameters that gives excellent results for closed-shell pi systems, has succeeded in using the half-electron model to calculate the heats of formation of conjugated radicals.[2]

Not only does the half-electron model allow the computation of energies of radicals, with the same parameters employed for closed-shell pi systems, it also has the simplicity of a closed-shell SCF treatment. The equations for the elements of the $\mathfrak{F}$ matrix in the half-electron model are given by (36) and (37) of Chapter 8; but in computing the bond orders, cognizance must be taken of the fact that one orbital contains half-electrons which contribute only half as much to the bond orders as the electrons in the other orbitals. In addition, the expression for the total energy, (39) of Chapter 8, must be modified by subtracting $\frac{1}{4}J_{i'i'}$ for the spurious repulsion between the fictitious half-electrons.

Problem 9.1. Calculate the $\mathfrak{F}$ matrix for the half-electron model of the allyl radical, using the parameters $\beta = -1.6$ eV, $\gamma_{11} = 11.1$ eV, $\gamma_{12} = 7.5$ eV and $\gamma_{13} = 5.3$ eV. Transform to a symmetry basis set, and find the MO's starting with Hückel orbitals. Does the spurious term $\frac{1}{4}J_{i'i'}$ in the potential for the half-filled orbital of the radical cause any error in the wavefunction for this MO in allyl? Why are the half-electron SCF MO's the same as the Hückel orbitals? Compute the total energy of the pi system of the allyl radical.

Different Orbitals for Different Spins

Although the half-electron model is convenient to use and gives good results in computing energies, it obviously will not give the correct wavefunctions for radicals, since it averages the potentials in which electrons of α and β spin move We have seen that in a radical the potentials for electrons of opposite spin are different, so that to obtain correct

[2] *J. Amer. Chem. Soc.*, **90**, 1953 (1968).

wavefunctions we must allow electrons of opposite spin to have different wavefunctions. The SCF method by which we will find the correct wavefunctions is known as unrestricted Hartree-Fock, since in it we no longer require two electrons of opposite spin to occupy the same MO.

In allowing electrons of opposite spin to occupy different orbitals, we are faced with the problem of having to solve two different sets of Fock equations, one for the electrons of α spin and one for those with spin β. Moreover, we may anticipate that the two sets of equations will be coupled, since the distribution of the α spin electrons will affect the wavefunctions for the β spin electrons and vice versa. Thus, in an unrestricted Hartree-Fock calculation, our self-consistent procedures become a little more complicated. We begin as usual with a set of assumed wavefunctions and use them to calculate a new set of wavefunctions for the electrons of α spin. However, before employing these MO's to calculate an even better set for the α spin electrons, we must first also calculate a new set of wavefunctions for the electrons of β spin. Both new sets of wavefunctions are then used to obtain another set of wavefunctions, and the process is repeated until self-consistency is reached.

Let us set up the equations for carrying out such a procedure to obtain a DODS wavefunction for the allyl radical. We must begin by obtaining expressions for the elements of the separate $\mathfrak{F}$ matrices for the electrons of α and of β spin. Rather than deriving these expressions mathematically, let us start with the equations for the $\mathfrak{F}$ matrix elements for a closed-shell molecule. Using the physical interpretation of what each term in them represents, we will see how they should be modified for a DODS calculation, where we explicitly allow for the fact that the effect of an electron on the potential in which another moves depends on whether their spin functions are the same or different. The diagonal elements of the $\mathfrak{F}$ matrix for a closed-shell molecule are, from Equation (36) of Chapter 8,

$$F_{rr} = -I_r + \tfrac{1}{2}P_{rr}\gamma_{rr} + \sum_{s \neq r}(P_{ss} - Z'_s)\gamma_{rs}$$

For an electron of α spin the factor $\tfrac{1}{2}P_{rr}$ in the second term gives the number of electrons of β spin that will simultaneously be at atom r, since another electron of α spin cannot simultaneously appear there. We therefore must replace $\tfrac{1}{2}P_{rr}$ by P_{rr}^{β} in our expression for F_{rr}^{α}. P_{ss} is carried over to our expression for F_{rr}^{α} unchanged, because P_{ss} is the net electron density on atom s, which is comprised of electrons of both α and β spin. In

Equation (37) of Chapter 8 for the off-diagonal elements,

$$F_{qr} = \beta_{qr} - \tfrac{1}{2} P_{qr} \gamma_{qr}$$

we saw that the term $-\tfrac{1}{2} P_{qr} \gamma_{qr}$ appears as a correction for the correlation of the motions of electrons of like spin. Thus, the factor $\tfrac{1}{2} P_{qr}$ should be replaced by P_{qr}^{α} in our expression for F_{qr}^{α}. The correct expressions for the elements of the $\mathfrak{F}$ matrix for electrons of α spin in a DODS calculation are then

$$F_{rr}^{\alpha} = -I_r + P_{rr}^{\beta}\gamma_{rr} + \sum_{s \neq r} (P_{ss} - Z_s')\gamma_{rs} \tag{6}$$

$$F_{qr}^{\alpha} = \beta_{qr} - P_{qr}^{\alpha}\gamma_{qr} \tag{7}$$

The expressions for the elements of $\mathfrak{F}^{\beta}$ can be obtained from (6) and (7) by just interchanging the roles of α and β.

The $\mathfrak{F}$ matrix for the electrons of α spin in allyl, starting with HMO's and assuming that the unpaired electron has α spin, is

$$[F^{\alpha}] = \begin{bmatrix} \tfrac{1}{4}\gamma_{11} & \beta - \tfrac{1}{4}\sqrt{2}\,\gamma_{12} & \tfrac{1}{4}\gamma_{13} \\ \beta - \tfrac{1}{4}\sqrt{2}\,\gamma_{12} & \tfrac{1}{2}\gamma_{11} & \beta - \tfrac{1}{4}\sqrt{2}\,\gamma_{12} \\ \tfrac{1}{4}\gamma_{13} & \beta - \tfrac{1}{4}\sqrt{2}\,\gamma_{12} & \tfrac{1}{4}\gamma_{11} \end{bmatrix} \tag{8}$$

We can see immediately that the larger diagonal element for atom 2 will tend to move the α electrons onto the terminal atoms of the allylic system, resulting in an increase in α density at these carbons.

Problem 9.2. Set up the $\mathfrak{F}$ matrix for the electron of β spin in allyl, using the Hückel MO's. Show that the diagonal elements of this matrix tend to move the β spin electron onto the central carbon atom. Do you expect this effect to be enhanced or lessened when the HMO wavefunctions for the α electrons are replaced by those obtained from diagonalization of the matrix $[F^{\alpha}]$ in (8)? What keeps the β spin electron from being confined exclusively to the central carbon atom?

Problem 9.3. Use the parameters of Problem 9.1 in a DODS computation of the MO's for the allyl radical. When the calculation has reached self-consistency, compare your calculated spin distribution with the experimental one of $+ 0.58$ at the terminal atoms and -0.16 at the central carbon.

Spin Operators and Their Eigenfunctions

From a DODS SCF treatment of allyl we obtain a wavefunction of the form

$$\Psi = |\psi_1^\alpha \psi_{1'}^\beta \psi_2^\alpha\rangle \qquad (9)$$

where the prime indicates that the MO's occupied by the electrons of α and β spin are not the same. Since the spin operators S_z and S^2 commute with the Hamiltonian for allyl, a proper wavefunction should be an eigenfunction of both these operators. We will now develop the mathematics of spin operators necessary to show that, although the wavefunction in (9) is an eigenfunction of S_z, it is not an eigenfunction of S^2 and so cannot be a legitimate wavefunction for the allyl radical.[3]

Spin operators belong to a more general class called angular momentum operators. A great deal can be derived about the properties of such operators solely from the fact that their components do not commute

$$L_x L_y \neq L_y L_x \qquad (10)$$

In fact,

$$L_x L_y - L_y L_x \equiv [L_x, L_y] = i\hbar L_z$$

$$L_y L_z - L_z L_y \equiv [L_y, L_z] = i\hbar L_x \qquad (11)$$

$$L_z L_x - L_x L_z \equiv [L_z, L_x] = i\hbar L_y$$

where $[L_x, L_y]$ is known as the commutator of L_x and L_y, and $\hbar$ is Planck's constant divided by 2π. It immediately follows from the definition of the commutator that

$$[L_x, L_y] = -[L_y, L_x] \qquad (12)$$

---

[3] Since the wavefunction obtained from a DODS SCF calculation cannot be the true wavefunction of the allyl radical, it is not surprising that such a wavefunction gives spin densities in poor agreement with experiment (see Problem 9.3).

Problem 9.4.   Use the commutation rules to show that L_z commutes with $L^2 = L_x^2 + L_y^2 + L_z^2$. Note that $[L_z, L_x^2] = [L_z, L_x] L_x + L_x [L_z, L_x]$, since the components of **L** behave like differential operators.

Problem 9.5.   Angular momentum is defined in classical mechanics as $L = r \times p$, so that the components of **L** are given by

$$L_x = yp_z - zp_y$$

$$L_y = zp_x - xp_z \qquad (13)$$

$$L_z = xp_y - yp_x$$

Show that in quantum mechanics replacing the components (p_x, p_y, p_z) of **p** by $-i\hbar(\partial/\partial x, \partial/\partial y, \partial/\partial z)$ gives a set of operators which satisfy the commutation rules.

It is possible to find the eigenvalues of angular momentum operators from their commutation relationships without ever specifying the exact form of the operators or their eigenfunctions. Let us define a new operator

$$L^+ \equiv L_x + iL_y \qquad (14)$$

The reason for the plus sign will become apparent shortly, for we are going to show that this operator raises the eigenvalue of an eigenfunction of L_z by one unit of $\hbar$. Let us represent an eigenfunction of L_z with eigenvalue g by the ket $|f, g\rangle$. The letter f denotes the ket's eigenvalue with respect to L^2; for since L_z and L^2 commute (Problem 9.4), the ket must also be an eigenfunction of the latter operator. Let us operate on this ket with L^+ and by operating with L_z on the resulting function, determine the effect of L^+ on the eigenvalue of L_z.

$$L_z L^+ |f, g\rangle = (L_z L_x + iL_z L_y)|f, g\rangle \qquad (15)$$

We can use the commutation relationships to bring L_z past L_x and L_y in the right-hand side of (15). Substituting for $L_z L_x$ and $L_z L_y$ from (11) we obtain

$$L_z L^+ |f, g\rangle = [L_x L_z + i\hbar L_y + i(L_y L_z - i\hbar L_x)]|f, g\rangle$$
$$= [(L_x + iL_y)L_z + \hbar(iL_y + L_x)]|f, g\rangle = L^+(L_z + \hbar)|f, g\rangle \qquad (16)$$

Carrying out the operation with L_z in the right-hand side of (16), we finally find

$$L_z L^+ | f, g \rangle = L^+(g + \hbar) | f, g \rangle = (g + \hbar) L^+ | f, g \rangle \qquad (17)$$

so it appears that $(L^+ | f, g \rangle)$ is an eigenfunction of L_z with eigenvalue $g + \hbar$. Thus, the effect of L^+ can be written as

$$L^+ | f, g \rangle = K | f, g + \hbar \rangle \qquad (18)$$

where K is a constant, since we cannot be certain that the new ket is normalized. We can be sure, however, that L^+ leaves the eigenvalue of L^2 unaffected, because these two operators commute; the proof is the same as that in Problem 9.4.

Problem 9.6.   Demonstrate that the effect of $L^- = L_x - iL_y$ is

$$L^- | f, g \rangle = K' | f, g - \hbar \rangle \qquad (19)$$

Now it might appear that we could go on raising the L_z eigenvalue, g, of a wavefunction by successive applications of L^+ or lowering it with L^- indefinitely, but this is not the case. We can demonstrate this by examining the integrals

$$
\begin{aligned}
\langle f, g | L^- L^+ | f, g \rangle &= \langle f, g | (L_x - iL_y)(L_x + iL_y) | f, g \rangle \\
&= \langle f, g | L_x^2 + L_y^2 + i(L_x L_y - L_y L_x) | f, g \rangle \\
&= \langle f, g | L_x^2 + L_y^2 - \hbar L_z | f, g \rangle = \langle f, g | L^2 - L_z^2 - \hbar L_z | f, g \rangle \\
&= (f - g^2 - \hbar g)\langle f, g | f, g \rangle = (f - g^2 - \hbar g)
\end{aligned} \qquad (20)
$$

and

$$\langle f, g | L^+ L^- | f, g \rangle = \langle f, g | L^2 - L_z^2 + \hbar L_z | f, g \rangle = (f - g^2 + \hbar g) \quad (21)$$

Now both of these must be positive because they are integrals over an operator multiplied by its complex conjugate. Therefore, we deduce

$$f \geq g^2 + \hbar |g| \qquad (22)$$

Thus, there must be maximum and minimum allowed eigenvalues g, and if we apply L^+ to the ket with the maximum eigenvalue, g_{max}, or L to the ket with the minimum eigenvalue, g_{min}, we must obtain zero.

$$L^+ |f, g_{max}\rangle = 0 \qquad L^- |f, g_{min}\rangle = 0 \qquad (23)$$

Operating on the first equation in (23) with L^- and on the second with L^+, multiplying each by the bra corresponding to the ket, and integrating, we get, using the results of (20) and (21),

$$\langle f, g_{max} | L^- L^+ | f, g_{max}\rangle = f - g_{max}^2 - \hbar g_{max} = 0 \qquad (24)$$

$$\langle f, g_{min} | L^+ L^- | f, g_{min}\rangle = f - g_{min}^2 + \hbar g_{min} = 0 \qquad (25)$$

Because both equations are equal to zero, they must be equal to each other, which implies that $g_{max} = -g_{min}$. Since we can get from g_{max} to g_{min} by operating with L^-, we know that an integral number, n, units of $\hbar$ separate the two. Thus,

$$g_{max} - g_{min} = 2g_{max} = n\hbar \qquad (26)$$

from which we conclude that the eigenvalues of L_z are integral or half-integral multiples of $\hbar$, depending on whether n is even or odd. Experimentally, it is found that spin angular momentum has eigenvalues which are half-integral multiples of $\hbar$. Using the fact that

$$g_{max} = \ell\hbar \qquad (27)$$

where ℓ is positive and an integer for orbital angular momentum and a half-integer for spin, we find from (24) that

$$f = \ell^2\hbar^2 + \ell\hbar^2 = \ell(\ell + 1)\hbar^2 \qquad (28)$$

It is conventional to label the kets which represent our wavefunctions by the integers or half-integers which appear in their eigenvalues, rather than by the eigenvalues themselves. Thus,

$$L^2 |\ell, m\rangle = \ell(\ell + 1)\hbar^2 |\ell, m\rangle \qquad (29)$$

$$L_z |\ell, m\rangle = m\hbar |\ell, m\rangle \qquad (30)$$

and (22) and (27) imply that m is restricted to the values $-\ell, -\ell + 1,$ $\ldots \ell - 1, \ell$

Problem 9.7. The ket $L^+ | f, g\rangle$ has as its corresponding bra $\langle f, g | L^-$. Use this fact to show that the correct normalization for the wavefunctions resulting from the shift operations in (18) and (19) is

$$L^+ |\ell, m\rangle = \hbar \sqrt{(\ell - m)(\ell + m + 1)} |\ell, m + 1\rangle$$

$$L^- |\ell, m\rangle = \hbar \sqrt{(\ell + m)(\ell - m + 1)} |\ell, m - 1\rangle$$

(31)

We have now derived the properties of generalized angular momentum operators and their eigenfunctions, solely from the commutation rules for the operators. To differentiate spin from orbital angular momentum, it is conventional to replace $L^2, L_z,$ and $L^\pm$ by $S^2, S_z,$ and $S^\pm$, and ℓ and m by s and m_s in (29) — (31). For a single electron it is also conventional to use $|\alpha\rangle$ and $|\beta\rangle$ instead of the kets $|\frac{1}{2}, \frac{1}{2}\rangle$ and $|\frac{1}{2}, -\frac{1}{2}\rangle$ to represent the two possible spin wavefunctions. The effects of the spin operators on the two wavefunctions for a single electron are displayed in the following equations

$$S^2 |\alpha\rangle = \tfrac{1}{2}(1 + \tfrac{1}{2}\hbar)^2 |\alpha\rangle = \tfrac{3}{4}\hbar^2 |\alpha\rangle \qquad S^2 |\beta\rangle = \tfrac{3}{4}\hbar^2 |\beta\rangle \quad (32)$$

$$S_z |\alpha\rangle = \tfrac{1}{2}\hbar |\alpha\rangle \qquad S_z |\beta\rangle = -\tfrac{1}{2}\hbar |\beta\rangle \quad (33)$$

$$S^+ |\alpha\rangle = 0 \quad S^- |\alpha\rangle = \hbar |\beta\rangle \quad S^+ |\beta\rangle = \hbar |\alpha\rangle \quad S^- |\beta\rangle = 0 \quad (34)$$

We are now ready to see why the DODS-SCF wavefunction (9) is not an eigenfunction of S^2. From (20) and (21) we can obtain a convenient expression for S^2 as

$$S^2 = \tfrac{1}{2}(S^+S^- + S^-S^+) + S_z^2 \tag{35}$$

where $S^+, S^-,$ and S_z are each a sum of one-electron operators. Let us write out S^2 for a two-electron case.

$$\begin{aligned}
S^2 = &\tfrac{1}{2}\{[S^+(1) + S^+(2)][S^-(1) + S^-(2)] + [S^-(1) + S^-(2)][S^+(1) + S^+(2)]\} \\
&+ [S_z(1) + S_z(2)]^2 \\
= &\tfrac{1}{2}\{S^+(1)S^-(1) + S^+(1)S^-(2) + S^+(2)S^-(1) + S^+(2)S^-(2) \\
&+ S^-(1)S^+(1) + S^-(1)S^+(2) + S^-(2)S^+(1) + S^-(2)S^+(2)\} \\
&+ [S_z(1) + S_z(2)]^2
\end{aligned} \tag{36}$$

The last term in brackets is the easiest to deal with since it just gives the square of the net m_s value for a wavefunction times $\hbar^2$ times the original wavefunction. The effects of the terms involving the raising and lowering operators are somewhat more complicated. The terms where both S^+ and S^- operate on the same electron give back the original wavefunction times $\hbar^2$ for each of the electrons, since

$$S^+S^-|\alpha\rangle = \hbar S^+|\beta\rangle = \hbar^2|\alpha\rangle \tag{37}$$

while $S^-S^+|\alpha\rangle$ gives zero. The roles of S^+S^- and S^-S^+ are reversed for an electron of β spin. The terms where S^+ and S^- operate on different electrons give zero if the electrons have the same spin. If the electrons have opposite spins, two of the terms still give zero, but the other two terms have the effect of interchanging the spins and multiplying the resulting wavefunction by $\hbar^2$, as the reader may verify for himself. Since this is also the case for each pair of electrons in a many-electron wavefunction, we can generalize our results and write the effect of S^2 on any wavefunction as

$$S^2\Psi = \hbar^2[\tfrac{1}{4}(n_\alpha - n_\beta)^2 + \tfrac{1}{2}n + \sum_{i\neq j} P_{ij}]\Psi \tag{38}$$

where n is the total number of electrons, $(n_\alpha - n_\beta)$ is the difference in number of those with α spin and β spin, and P_{ij} is an operator which interchanges the spins of electrons i and j, *provided* that their spins are *different*.

Problem 9.8. Use (38) to show that the wavefunctions $|\psi_1^\alpha\psi_1^\beta\rangle$ and $|\psi_1^\alpha\psi_2^\alpha\rangle$ are eigenfunctions of S^2. Show that, in contrast, $|\psi_1^\alpha\psi_2^\beta\rangle$ is not. Confirm that the $m_s = 0$ wavefunctions for the excited states of ethylene, $\Psi = 1/\sqrt{2}(\langle|\psi_1^\alpha\psi_2^\beta\rangle - |\psi_1^\beta\psi_2^\alpha\rangle)$ and $\Psi = 1/\sqrt{2}(\langle|\psi_1^\alpha\psi_2^\beta\rangle + |\psi_1^\beta\psi_2^\alpha\rangle)$, are respectively the singlet and triplet.

From Equation (38) we can see that not every wavefunction constructed from a single Slater determinant will be an eigenfunction of S^2, because the operator P_{ij}, which interchanges different electron spin wavefunctions, can generate other wavefunctions that are not just multiples of the original. In fact, there are only three cases where a single Slater determinant will be an eigenfunction of S^2. The first occurs when all the electrons have the same spin. Because P_{ij} does not interchange electrons of the same spin, such a wavefunction is obviously an eigenfunction of S^2 with eigenvalue $\frac{1}{2}n (\frac{1}{2}n + 1)\hbar^2$. The second type of Slater determinant which, by itself, is an eigenfunction of S^2 is one which represents a closed-shell system in which all the MO's are doubly filled. The reason that such a wavefunction is an eigenfunction of S^2 is that when P_{ij} interchanges electron spins between different orbitals, an electron of β spin is removed from an MO and replaced by one of α spin. However, the electron which remains in that orbital also has α spin, and so the wavefunction vanishes. For an interchange of spin functions between two electrons in the same MO, the wavefunction does not vanish; but it is the same as the initial wavefunction, except for the exchange of two rows in the Slater determinant. The new wavefunction is, therefore, just the negative of the original wavefunction; and since there are $n/2$ MO's within which the spin functions can be interchanged, (38) informs us of the unsurprising result that the eigenvalue of S^2 for a closed-shell is $\hbar^2(\frac{1}{2}n - n/2) = 0$. The third case in which a single Slater determinant is an eigenfunction of S^2 is really a combination of the first two and occurs when all the electrons outside a closed-shell have the same spin.

Since the Hückel wavefunction for allyl belongs to case three, it is an eigenfunction of S^2. However, the DODS wavefunction is not an eigenfunction of S^2, because ψ_1 and ψ'_1 are different MO's. In fact,

$$S^2 |\psi_1^\alpha \psi_1^\beta \psi_2^\alpha\rangle = \hbar^2[(\frac{3}{2} + \frac{1}{4})|\psi_1^\alpha \psi_1^\beta \psi_2^\alpha\rangle + |\psi_1^\beta \psi_1^\alpha \psi_2^\alpha\rangle + |\psi_1^\alpha \psi_1^\alpha \psi_2^\beta\rangle]$$

(39)

The reason that the DODS wavefunction (9), composed of the single Slater determinant $|\psi_1^\alpha \psi_1^\beta \psi_2^\alpha\rangle$, is not an eigenfunction of S^2 is that it contains an admixture of the $m_s = \frac{1}{2}$ component of a quartet state. We can verify this by obtaining this component of the quartet. We know that $|\psi_1^\alpha \psi_1^\alpha \psi_2^\alpha\rangle$ is the quartet component with $m_s = \frac{3}{2}$, that is $|\frac{3}{2}, \frac{3}{2}\rangle$. We can obtain the component $|\frac{3}{2}, \frac{1}{2}\rangle$ by using S^-, since

$$S^- |\frac{3}{2}, \frac{3}{2}\rangle = S^- |\psi_1^\alpha \psi_1^\alpha \psi_2^\alpha\rangle = \hbar\sqrt{(\frac{3}{2} + \frac{3}{2})(\frac{3}{2} - \frac{3}{2} + 1)}|\frac{3}{2}, \frac{1}{2}\rangle$$

(40)

Carrying out the lowering operations we obtain

$$|\frac{3}{2}, \frac{1}{2}\rangle = \frac{1}{\sqrt{3}}(|\psi_1^\beta \psi_1^\alpha \psi_2^\alpha\rangle + |\psi_1^\alpha \psi_1^\beta \psi_2^\alpha\rangle + |\psi_1^\alpha \psi_1^\alpha \psi_2^\beta\rangle)$$

(41)

which is clearly not orthogonal to (9) and so must be contained in it. In fact, since there are a total of three Slater determinants with $m_s = \frac{1}{2}$, there must be three states which have components with this eigenvalue of S_z. One of them is the component of the quartet that we have already found. Since there are no states of higher multiplicity possible for three electrons than a quartet, the other two states must be doublets. Thus, our single Slater determinant DODS wavefunction (9) must be a linear combination of these three states; and to make it an eigenfunction of S^2, we must eliminate the component of the quartet from it. We can do this in a number of ways, perhaps the most straightforward being through the use of a doublet projection operator to annihilate the quartet contaminant.

Projection and Annihilation Operators

We have seen that our DODS wavefunction (9) is really a linear combination of several states, two doublets and a quartet.

$$\Psi^{\text{DODS}} = c_1 \Psi^{\text{IV}} + c_2 \Psi_a^{\text{II}} + c_3 \Psi_b^{\text{II}} \qquad (42)$$

Thus, when we operate on our DODS wavefunction with S^2 we obtain

$$S^2 \Psi^{\text{DODS}} = \tfrac{3}{2}(\tfrac{3}{2} + 1)\hbar^2 c_1 \Psi^{\text{IV}} + \tfrac{1}{2}(\tfrac{1}{2} + 1)\hbar^2 (c_2 \Psi_a^{\text{II}} + c_3 \Psi_b^{\text{II}}) \qquad (43)$$

The component of the quartet can be eliminated from the resulting function by subtracting $\tfrac{3}{2}(\tfrac{3}{2} + 1)\hbar^2\, c_1 \Psi^{\text{IV}}$ from it. This can be accomplished by operating on Ψ^{DODS} with $S^2 - \tfrac{3}{2}(\tfrac{3}{2} + 1)\hbar^2$. This operator annihilates the component of a wavefunction with $s = \tfrac{3}{2}$, and operating on (42) with it we obtain

$$[S^2 - \tfrac{3}{2}(\tfrac{3}{2} + 1)\hbar^2]\Psi^{\text{DODS}} = [\tfrac{1}{2}(\tfrac{1}{2} + 1)\hbar^2 - \tfrac{3}{2}(\tfrac{3}{2} + 1)\hbar^2][c_2 \Psi_a^{\text{II}} + c_3 \Psi_b^{\text{II}}]$$
$$(44)$$

We can eliminate the factor in the first bracket by defining our projection operator for obtaining a pure doublet wavefunction as[4]

---

[4] In the event that a spin function is contaminated with several other multiplicities, the projection operator is a product of annihilation operators, one for each contaminant. The general form of a projection operator that gives a component with spin quantum number s is

$$P = \prod_{s'} \frac{S^2 - s'(s' + 1)\hbar^2}{[s(s + 1) - s'(s' + 1)]\hbar^2}$$

where s' is the spin quantum number of each contaminant.

$$P^{\mathrm{II}} = \frac{S^2 - \frac{3}{2}(\frac{3}{2} + 1)\hbar^2}{[\frac{1}{2}(\frac{1}{2} + 1) - \frac{3}{2}(\frac{3}{2} + 1)]\hbar^2} = \frac{S^2 - \frac{15}{4}\hbar^2}{-3\hbar^2} \qquad (45)$$

Employing this operator and making use of (39) we obtain our projected DODS wavefunction after normalization as

$$P^{\mathrm{II}}\Psi^{\mathrm{DODS}} = \frac{1}{\sqrt{6}}(2\,|\,\psi_1^\alpha\psi_1^\beta\cdot\psi_2^\alpha\rangle - |\,\psi_1^\beta\psi_1^\alpha\cdot\psi_2^\alpha\rangle - |\,\psi_1^\alpha\psi_1^\alpha\cdot\psi_2^\beta\rangle) \qquad (46)$$

There is now one more problem which confronts us. Although we have obtained a doublet wavefunction which we projected from the best possible single determinantal DODS-SCF wavefunction, we have no guarantee that it is now the best possible doublet. In fact, it is not, since it is easy to show that it can be improved by configuration interaction with another doublet, which can be found by constructing a wavefunction that is orthogonal to both (46) and the quartet (41).

Problem 9.9.   Find the other doublet and show that (46) can be improved through mixing with it by demonstrating that there exists a nonzero matrix element of the two-electron operator between the two doublet wavefunctions.

Despite the fact that configuration interaction with another doublet can, in theory, improve the projected DODS wavefunction, the mixing is, in practice, expected to be small. The largest contribution to the projected DODS wavefunction is made by the ket that was optimized to be the single determinantal wavefunction of lowest possible energy. This ket is totally absent from the wavefunction for the other doublet; therefore, the second doublet may be anticipated to be of considerably higher energy than the one projected from the DODS wavefunction. Consequently, the latter should be a good approximation to the true wavefunction. Indeed, Amos and Snyder[5] have found that the projected DODS wavefunctions for a number of conjugated radicals give spin densities that are in excellent agreement with experiment. In addition, Amos and Snyder found that in systems larger than allyl, where multiplets higher than

[5] *J. Chem. Phys.* **41**, 1773 (1964) and **42**, 3670 (1965)

quartets can contaminate the unprojected DODS wavefunction, it is unnecessary to annihilate these higher multiplets in the DODS wavefunction in order to obtain good agreement with experiment. Multiplicities higher than the quartet require electrons to occupy MO's of very high energy. Therefore, these higher multiplicities contribute so little to the optimized, unprojected, DODS wavefunction that their small contributions do not need to be annihilated.

The arguments, used to rationalize why it is only necessary to annihilate the quartet contaminants in a DODS wavefunction for a radical and why it is possible to ignore mixing of the resulting wavefunction with other doublets, are similar to those given to justify considering mixing only between a limited number of configurations in improving electron correlation by CI. We now turn to consideration of this alternative approach to the calculation of spin densities in radicals.

Configuration Interaction Treatment
of Open-Shell Systems

Let us carry out a CI treatment of the allyl radical, starting with HMO's. The configuration of lowest energy, $\Psi_1 = |\psi_1^{\alpha}\psi_1^{\beta}\psi_2^{\alpha}\rangle$, has $m_s = \frac{1}{2}$ and is of the same symmetry as ψ_2. The following configurations also have this value of m_s and the same symmetry: $|\psi_1^{\alpha}\psi_3^{\beta}\psi_2^{\alpha}\rangle$, $|\psi_1^{\beta}\psi_3^{\alpha}\psi_2^{\alpha}\rangle$, $|\psi_1^{\alpha}\psi_3^{\beta}\psi_2^{\beta}\rangle$, and $|\psi_3^{\alpha}\psi_3^{\beta}\psi_2^{\alpha}\rangle$. The fourth of these has considerably higher energy than the first three, and so we shall ignore it for the present. We could now set up a 4×4 matrix representing the interactions between the three excited configurations and the lowest one. The diagonalization of this matrix would give us the energy and wavefunction for the ground state. However, we know that the lowest energy configuration has the quantum number $s = \frac{1}{2}$ for the S^2 operator; therefore, it will only mix with other functions which are doublets. We can use this fact to avoid diagonalizing the 4×4 matrix by constructing from the three excited configurations linear combinations that are eigenfunctions of S^2 with spin quantum number $s = \frac{1}{2}$. From the results of the previous section we know that there are two such functions, and one possible choice for them is $\Psi_2 = 1/\sqrt{6}\,(2|\psi_1^{\alpha}\psi_3^{\alpha}\psi_2^{\beta}\rangle - |\psi_1^{\beta}\psi_3^{\alpha}\psi_2^{\alpha}\rangle - |\psi_1^{\alpha}\psi_3^{\beta}\psi_2^{\alpha}\rangle)$ and $\Psi_3 = 1/\sqrt{2}\,(|\psi_1^{\alpha}\psi_3^{\beta}\psi_2^{\alpha}\rangle - |\psi_1^{\beta}\psi_3^{\alpha}\psi_2^{\alpha}\rangle)$. Although any linear combination of these two is also an eigenfunction of S^2 with $s = \frac{1}{2}$, this set proves to be a particularly good choice because Ψ_2 and Ψ_3 are not mixed by the Hamiltonian.[6]

Problem 9.10. Verify that $\langle \Psi_2 | \mathcal{H} | \Psi_3 \rangle = 0$. [Hint: Show that $(\psi_1\psi_2 | \psi_1\psi_2) = (\psi_3\psi_2 | \psi_3\psi_2)$.]

We now proceed to compute the matrix elements between the lowest energy configuration and the excited doublets by calculating the interaction between the former and each ket in the wavefunctions of the latter. First, we note that $|\psi_1^\alpha \psi_1^\beta \psi_2^\alpha\rangle$ differs from $|\psi_1^\beta \psi_3^\alpha \psi_2^\alpha\rangle$ and $|\psi_1^\alpha \psi_3^\beta \psi_2^\alpha\rangle$ in the orbital assignment of only one electron; therefore, interaction between the former and the latter two configurations can occur through a one- as well as the two-electron operator.

$$\langle \psi_1^\alpha \psi_1^\beta \psi_2^\alpha | \hbar | \psi_1^\alpha \psi_3^\beta \psi_2^\alpha \rangle = \langle \psi_1^\alpha \psi_2^\alpha | \psi_1^\alpha \psi_2^\alpha \rangle \langle \psi_1^\beta | \hbar | \psi_3^\beta \rangle = \langle \psi_1 | \hbar | \psi_3 \rangle \quad (47)$$

Since the MO's are Hückel orbitals, any nonzero integral that arises from (47) must come from a one-electron operator that is not present in the Hückel Hamiltonian. The only candidate is the nuclear-electron attraction operator, and the nuclei do, indeed, have a net interaction with the electron distribution given by

$$\psi_1\psi_3 = \tfrac{1}{4}(\phi_1^2 + \phi_3^2) - \tfrac{1}{2}\phi_2^2 \quad (48)$$

as shown schematically in Figure 9.1. (Remember that electrons have a charge of $-e$.)

---

[6] Two other sets of doublet wavefunctions can be generated by using the doublet projection operator on each of the three excited configurations and taking linear combinations of the resulting functions to achieve orthogonality. However, the other sets of doublets are mixed by the Hamiltonian to give Ψ_2 and Ψ_3. The reader may have noted the similarity between the spin wavefunctions for doublets and the spatial wavefunctions for the degenerate pair of orbitals in molecules of C_{3v} or D_3 symmetry. The resemblance is not coincidental, for each operation in the permutation group to which a three spin system belongs corresponds to one in these point groups. Therefore, the representations of these groups are identical, and the groups are said to be isomorphous. Permutation groups can be used to simplify problems involving spin wavefunctions in much the same way that point groups can be used for spatial wavefunctions. For instance, the nonmixing of Ψ_2 and Ψ_3 is not accidental but can be anticipated on the basis of the fact that permutation of the spins of the electrons in ψ_1 and ψ_3 leaves the energy of the kets unchanged, so that this operation commutes with the Hamiltonian. Since Ψ_2 and Ψ_3 have different eigenvalues for this operation, they are not mixed by the Hamiltonian.

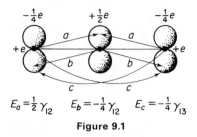

$$E_a = \tfrac{1}{2}\gamma_{12} \qquad E_b = -\tfrac{1}{4}\gamma_{12} \qquad E_c = -\tfrac{1}{4}\gamma_{13}$$

Figure 9.1

The interaction energy of this electron distribution with the nuclei can also be represented in our usual notation for electron repulsion integrals by expressing the effective nuclear charge at each atom as $-\phi_r^2$. Thus,

$$\langle \psi_1 | \hbar | \psi_3 \rangle = (\tfrac{1}{4}[\phi_1^2 + \phi_3^2] - \tfrac{1}{2}\phi_2^2 | - [\phi_1^2 + \phi_2^2 + \phi_3^2]) = \tfrac{1}{2}(\gamma_{12} - \gamma_{13}) \tag{49}$$

The integral $\langle \psi_1^\alpha \psi_1^\beta \psi_2^\alpha | \hbar | \psi_1^\beta \psi_3^\alpha \psi_2^\alpha \rangle$ has the same magnitude but the opposite sign.

$$\langle \psi_1^\alpha \psi_1^\beta \psi_2^\alpha | \hbar | \psi_1^\beta \psi_3^\alpha \psi_2^\alpha \rangle = -\langle \psi_1 | \hbar | \psi_3 \rangle = -\tfrac{1}{2}(\gamma_{12} - \gamma_{13}) \tag{50}$$

The two-electron matrix elements are

$$\left\langle \psi_1^\alpha \psi_1^\beta \psi_2^\alpha \left| \frac{e^2}{r_{12}} \right| \psi_1^\alpha \psi_3^\beta \psi_2^\alpha \right\rangle = \left\langle \psi_1^\alpha \psi_1^\beta \left| \frac{e^2}{r_{12}} \right| \psi_1^\alpha \psi_3^\beta \right\rangle + \left\langle \psi_1^\beta \psi_2^\alpha \left| \frac{e^2}{r_{12}} \right| \psi_3^\beta \psi_2^\alpha \right\rangle$$
$$= (\psi_1 \psi_1 | \psi_1 \psi_3) + (\psi_1 \psi_3 | \psi_2 \psi_2) \tag{51}$$

$$\left\langle \psi_1^\alpha \psi_1^\beta \psi_2^\alpha \left| \frac{e^2}{r_{12}} \right| \psi_1^\beta \psi_3^\alpha \psi_2^\alpha \right\rangle = -\left\langle \psi_1^\alpha \psi_1^\beta \left| \frac{e^2}{r_{12}} \right| \psi_3^\alpha \psi_1^\beta \right\rangle - \left\langle \psi_1^\alpha \psi_2^\alpha \left| \frac{e^2}{r_{12}} \right| \psi_3^\alpha \psi_2^\alpha \right\rangle$$
$$= -(\psi_1 \psi_3 | \psi_1 \psi_1) - (\psi_1 \psi_3 | \psi_2 \psi_2) \tag{52}$$
$$+ (\psi_1 \psi_2 | \psi_2 \psi_3)$$

$$\left\langle \psi_1^\alpha \psi_1^\beta \psi_2^\alpha \left| \frac{e^2}{r_{12}} \right| \psi_1^\alpha \psi_3^\alpha \psi_2^\beta \right\rangle = -\left\langle \psi_1^\beta \psi_2^\alpha \left| \frac{e^2}{r_{12}} \right| \psi_2^\beta \psi_3^\alpha \right\rangle = -(\psi_1 \psi_2 | \psi_2 \psi_3) \tag{53}$$

We can check our work by confirming that there is no interaction between Ψ_1 and the $m_s = \tfrac{1}{2}$ component of the quartet, since (51) − (53) sum to zero, as do (47) and (50). We are now able to compute

$$\langle \Psi_1 | \mathcal{H} | \Psi_2 \rangle = \frac{1}{\sqrt{6}}[-2(\psi_1\psi_2|\psi_2\psi_3) - (\psi_1\psi_2|\psi_2\psi_3)]$$

$$= -\tfrac{1}{2}\sqrt{6}\,(\psi_1\psi_2|\psi_2\psi_3)$$

(54)

and

$$\langle \Psi_1 | \mathcal{H} | \Psi_3 \rangle = \frac{1}{\sqrt{2}}(\gamma_{12} - \gamma_{13}) + \frac{1}{\sqrt{2}}[2(\psi_1\psi_3|\psi_1\psi_1)$$

$$+ 2(\psi_1\psi_3|\psi_2\psi_2) - (\psi_1\psi_2|\psi_2\psi_3)]$$

(55)

Problem 9.11.   Show that

$$\langle \Psi_1 | \mathcal{H} | \Psi_2 \rangle = -\tfrac{1}{2}\sqrt{6}\,K_{12} = -\tfrac{1}{8}\sqrt{6}\,(\gamma_{11} - \gamma_{13}) \qquad (56)$$

Calculate the integrals in (55) and demonstrate that Ψ_1 and Ψ_3 are not mixed by the Hamiltonian, as predicted by Brillouin's theorem.

In order to compute the mixing between Ψ_1 and Ψ_2 we must now compute the difference in their energies. The wavefunctions give the same gross electron distribution; therefore, they have identical Coulomb integrals in the expressions for their energies. However, while the energy of Ψ_1 contains the term $-K_{12}$, it can be shown that the expression for the energy of Ψ_2 contains $K_{12} - K_{13}$. In addition, Ψ_2, of course, has an electron in ψ_3 instead of in ψ_1, so its one-electron energy is $-2 \times \sqrt{2}\,\beta$ greater than that of Ψ_1. Evaluating K_{12} and K_{13}, the difference in energy between Ψ_1 and Ψ_2 is

$$E(\Psi_2) - E(\Psi_1) = -2\sqrt{2}\,\beta + 2K_{12} - K_{13}$$

$$= -2.8\beta + \tfrac{1}{8}(\gamma_{11} + 4\gamma_{12} - 5\gamma_{13})$$

(57)

The number $-2\sqrt{2}\,\beta$ in (57) represents the difference between the one-electron energy of a ground and an excited state; therefore, we should use a spectroscopic β to compute this quantity. From the bond order in the ground state of the allyl radical we can calculate, using (11) and (12) of Chapter 3 and the spectroscopic value of β in ethylene, that -2.5 eV is an appropriate value of β for use in (57). Substituting values of the repulsion integrals from Problem 9.1 into (56) and (57), the mixing coefficient of Ψ_2 into Ψ_1 can be calculated from perturbation theory as

$$c = \frac{\langle \Psi_1 | \mathcal{K} | \Psi_2 \rangle}{E(\Psi_1) - E(\Psi_2)} = \frac{-1.8}{-8.8} = +0.20 \qquad (58)$$

The mixing is calculated to lower the energy of the resulting wavefunction for the ground state of allyl by

$$\Delta E = \frac{\langle \Psi_1 | \mathcal{K} | \Psi_2 \rangle^2}{E(\Psi_1) - E(\Psi_2)} = -0.36 \, \text{eV} \qquad (59)$$

Since the mixing coefficient between Ψ_1 and Ψ_2 is quite small, the factor required to normalize the CI wavefunction is very close to unity (it is, in fact, 0.98); therefore, for computational simplicity in calculating the spin densities let us continue to use the unnormalized wavefunction

$$\Psi = \Psi_1 + 0.20\Psi_2 = |\psi_1^\alpha \psi_1^\beta \psi_2^\alpha \rangle + \frac{0.20}{\sqrt{6}} (2 |\psi_1^\alpha \psi_3^\alpha \psi_2^\beta \rangle$$
$$- |\psi_1^\beta \psi_3^\alpha \psi_2^\alpha \rangle - \psi_1^\alpha \psi_3^\beta \psi_2^\alpha \rangle) \qquad (60)$$

If we were to expand the MO's in the kets in this wavefunction, as we did for the CI ethylenic wavefunction in Chapter 7, we would find that the first term in Ψ_2 improves the correlation between the electron of β spin in ψ_1 and the unpaired electron in ψ_2 by increasing the probability that when one is on atom 1, the other will be found on atom 3. This term, however, has little effect on the spin distribution in allyl. The latter two terms in Ψ_2 improve the electron correlation by increasing the probability of finding the electron in ψ_1 with α spin at the terminal atoms and the electron of β spin at the central atom, as we have discussed before.[7] Thus, it

---

[7] Note that $|\psi_1^\alpha \psi_1^\beta \psi_2^\alpha \rangle - c|\psi_1^\beta \psi_3^\alpha \psi_2^\alpha \rangle$ can be rewritten as $|(\psi_1 + c\psi_3)^\alpha \psi_1^\beta \psi_2^\alpha \rangle$. Mixing ψ_3 into ψ_1 with a plus sign decreases the coefficient at the central atom and increases those at the terminal atoms for the α spin electron in ψ_1. In the same fashion it can be shown that $-c|\psi_1^\alpha \psi_3^\beta \psi_2^\alpha \rangle$ has just the opposite effect on the wavefunction for the β spin electron in ψ_1; for this term mixes ψ_3 into ψ_1 with a minus sign. Mixing of the other excited doublet, Ψ_3, into Ψ_1 would produce identical changes in the wavefunctions for both electrons in ψ_1 with resulting net separation of charge within the molecule. Such mixing could, therefore, only raise the energy of the wavefunction for the ground state, and so it does not occur. A similar type of analysis, based on MO mixing, can be used to establish the mechanism by which the first term in Ψ_2 improves electron correlation in the allyl radical.

is predominantly the last two terms, which together are responsible for only 0.12 eV of the energy lowering, that give rise to the anomalous (from the point of view of simple Hückel theory) spin density in allyl.

We can verify that this is the case by computing the spin distribution given by (60), using the difference between the operators $\delta^\alpha(r)$ and $\delta^\beta(r)$, which were defined in Chapter 8, to obtain the net spin density at each atom. We wish to calculate at each atom, r,

$$
\begin{aligned}
\langle \Psi | \delta^\alpha(r) - \delta^\beta(r) | \Psi \rangle = {} & \langle \Psi_1 | \delta^\alpha(r) - \delta^\beta(r) | \Psi_1 \rangle \\
& + 0.40 \langle \Psi_1 | \delta^\alpha(r) - \delta^\beta(r) | \Psi_2 \rangle \\
& + 0.04 \langle \Psi_2 | \delta^\alpha(r) - \delta^\beta(r) | \Psi_2 \rangle
\end{aligned}
\tag{61}
$$

Since the first term just gives the spin distribution of the simple HMO wavefunction and since the last term, because of its coefficient, is much smaller than the second (see Problem 9.12), the chief contribution to the anomalous spin distribution comes from the cross term

$$
\begin{aligned}
\langle \Psi_1 | \delta^\alpha(r) - \delta^\beta(r) | \Psi_2 \rangle = {} & \frac{2}{\sqrt{6}} \langle \psi_1^\alpha \psi_1^\beta \psi_2^\alpha | \delta^\alpha(r) - \delta^\beta(r) | \psi_1^\alpha \psi_3^\alpha \psi_2^\beta \rangle \\
& - \frac{1}{\sqrt{6}} \langle \psi_1^\alpha \psi_1^\beta \psi_2^\alpha | \delta^\alpha(r) - \delta^\beta(r) | \psi_1^\beta \psi_3^\alpha \psi_2^\alpha \rangle \\
& - \frac{1}{\sqrt{6}} \langle \psi_1^\alpha \psi_1^\beta \psi_2^\alpha | \delta^\alpha(r) - \delta^\beta(r) | \psi_1^\alpha \psi_3^\beta \psi_2^\alpha \rangle
\end{aligned}
\tag{62}
$$

Since $\delta^\alpha(r) - \delta^\beta(r)$ is a one-electron operator, the first of the terms in (62) is zero, because $|\psi_1^\alpha \psi_1^\beta \psi_2^\alpha \rangle$ and $|\psi_1^\alpha \psi_3^\alpha \psi_2^\beta \rangle$ differ in two one-electron wavefunctions.

The remaining two terms reduce to

$$
\begin{aligned}
\langle \Psi_1 | \delta^\alpha(r) - \delta^\beta(r) | \Psi_2 \rangle = {} & \frac{1}{\sqrt{6}} [\langle \psi_1^\alpha | \delta^\alpha(r) - \delta^\beta(r) | \psi_3^\alpha \rangle \\
& - \langle \psi_1^\beta | \delta^\alpha(r) - \delta^\beta(r) | \psi_3^\beta \rangle]
\end{aligned}
\tag{63}
$$

When evaluated at atoms 1 or 3, Equation (63) gives

$$
\langle \Psi_1 | \delta^\alpha(1) - \delta^\beta(1) | \Psi_2 \rangle = \frac{1}{\sqrt{6}} \left[\left(\frac{1}{2} \times \frac{1}{2} \right) - \left(-\frac{1}{2} \times \frac{1}{2} \right) \right] = \frac{1}{2\sqrt{6}}
\tag{64}
$$

net α spin density. At atom 2

$$\langle \Psi_1 | \delta^\alpha(2) - \delta^\beta(2) | \Psi_2 \rangle = \frac{1}{\sqrt{6}}\left[-\frac{1}{\sqrt{2}} \times \frac{1}{\sqrt{2}} - \left(-\frac{1}{\sqrt{2}} \times -\frac{1}{\sqrt{2}} \right) \right]$$

$$= -\frac{1}{\sqrt{6}} \tag{65}$$

the minus sign indicating that there is net β spin density at atom 2. In Equation (61) for the spin densities, the term $\langle \Psi_1 | \delta^\alpha(r) - \delta^\beta(r) | \Psi_2 \rangle$ is multiplied by a factor of 0.40. Therefore, its contribution to the spin distribution, calculated from the CI wavefunction, is to produce an additional α spin density of $0.40/2\sqrt{6} = 0.082$ at the terminal atoms and a β spin density of 0.164 at the central carbon atom in allyl. These values are in excellent agreement with those calculated from the spectrum of the allyl radical, assuming Q in Equation (14) of Chapter 5 is equal to 24.7 gauss—the algebraic sum of the hyperfine coupling constants.

Problem 9.12. Show that the third term in (61) contributes net α spin density of 0.027 at the central and 0.007 at each of the terminal carbon atoms. Calculate the spin densities in the allyl radical, including all the terms in (61) and the correct normalization factor for the wavefunction.

Summary

From the point of view of energetics, wavefunctions that assign electrons of opposite spin to identical MO's are no worse in radicals than in closed-shell molecules, since in both types of systems the same kinds of correlation effects are ignored. If we are interested only in energetics, use of Dewar's half-electron method enables us to employ a closed-shell SCF approach to calculations on radicals. Dewar has found that in this method the same parameters employed for closed-shell molecules can be used. Since similar correlation effects are ignored in the wavefunctions for both open- and closed-shell systems, the same semiempirical parameters, which implicitly include these effects, are appropriate for use with both types of systems. In contrast, in a calculation on an open-shell system, where we explicitly allow for electron correlation in the wavefunction, we are being somewhat inconsistent if we use parameters for closed-shell molecules. In effect, we are then including correlation effects twice—once explicitly in our wavefunction and once implicitly in our parameters.

In spite of the convenience of using a closed-shell approach to open-shell systems, it is still often desirable to carry out calculations where cognizance is explicitly taken of the fact that electrons of opposite spin have different wavefunctions. The motivation is, of course, that in radicals the effects of this kind of electron correlation can be experimentally observed in the epr spectra.[8] In this chapter we saw how, within the SCF formalism, this kind of correlation can be built into the wavefunction by allowing different orbitals for different spins (DODS) and using separate $\mathfrak{F}$ matrices for α and β spin electrons. Our investigation of spin operators led us, however, to the conclusion that a single DODS Slater determinant is not an eigenfunction of S^2. It was, nevertheless, possible to obtain an eigenfunction of S^2 from this wavefunction by the use of a projection operator, which annihilates the quartet contaminant of a doublet wavefunction.

In a CI approach to improving the wavefunction for a radical, the result will automatically be an eigenfunction of S^2, because this operator commutes with the Hamiltonian. Nevertheless, we saw the utility of combining excited configurations into eigenfunctions of S^2 before carrying out the CI; in the case of allyl this procedure saved us the chore of diagonalizing a 4×4 matrix.

CI and spin projected DODS wavefunctions, when both are completely optimized, must ultimately prove to be identical,[9] despite the differences in the two approaches. However, we will rarely take the trouble to obtain the absolutely best possible wavefunction by either method. In doing CI we will usually limit ourselves to mixing in just a few excited configurations; for instance, we ignored $|\psi_3^\alpha \psi_3^\beta \psi_2^\alpha\rangle$ in our CI treatment of allyl. Similarly, in projecting our DODS wavefunctions, we will usually be content to leave unannihilated all but the next highest contaminating multiplet; and then we will ignore mixing of the (relatively pure) doublet that we obtain with other excited doublets. In deciding which approach

[8] Correlation effects can also be observed in the epr spectra of other open-shell systems; for instance, negative spin density at the central carbon has a large effect on the epr parameters of the triplet ground state of trimethylenemethane. Such molecules can also be treated by either DODS or CI methods.

[9] This is only true, of course, in semiempirical methods if the same set of parameters is used to obtain both wavefunctions.

to employ for a given problem, we must evaluate which method will give better results at the level of approximation at which we plan to use it.

The drawback to a partial CI is that we never know when to quit, especially in a complex molecule where there may be a large number of relatively low-lying excited configurations. In the DODS-SCF approach there is not this ambiguity; we iterate the SCF procedure until self-consistency is reached and then project the resulting Slater determinant. Although we can not say a priori whether the CI or projected DODS wavefunction will give calculated spin densities in closer agreement with experiment for a given radical, we can be relatively certain that the quality of the results obtained by the latter method will be more consistent (good or bad) than those of the former, since the effects due to subjective judgment of the theoretician are absent from the DODS wavefunction.

The advantage to a CI calculation is that a simple one, for instance on allyl, is easier to carry out by hand than the corresponding DODS-SCF computation. In addition, the spin densities are somewhat easier to calculate from a simple CI wavefunction; for the orbitals for α and β spin electrons are generally not orthogonal in a DODS-SCF wavefunction. Therefore, to obtain quickly a semiquantitative estimate of "anomalous" spin density distribution in the absence of a DODS-SCF computer program, one might wish to carry out by hand a CI calculation, mixing only one excited configuration into the lowest one.

Whatever the computational approach chosen, it is clear that both DODS-SCF and CI calculations on radicals lead to wavefunctions for electrons of opposite spin that tend to keep the electrons on different atoms so that their electrostatic repulsion is minimized. "Anomalous" spin densities are then seen as a natural result of the inclusion in the Hamiltonian of the electron repulsion operator.

FURTHER READING

LIONEL SALEM, *Molecular Orbital Theory of Conjugated Systems*, W. A. Benjamin, Inc., New York, 1966, Chapter 5 and references therein, particularly A. D. MCLACHLAN, *Mol. Phys.*, 3, 233 (1960).

10

All-Valence
Electron Calculations

The application to π systems of a theoretical method that explicitly includes electron repulsion in the Hamiltonian has been discussed in the preceding three chapters. Clearly, it is desirable to extend this method to the treatment of all-valence, rather than just π, electrons; and this extension forms part of the subject material for this chapter. In the past decade the theoretical treatment of sigma systems has received considerable attention, and three books and several review articles that deal with methods and applications have appeared. Rather than duplicating here all the material readily available from these sources, the aim of this chapter is merely to provide a brief survey of the three main types of theoretical approaches to all-valence electron calculations, the approximations they contain, and some of their applications. More extensive information is available from the literature listed at the end of this chapter.

Extended Hückel Theory (EHT)

Roald Hoffmann in 1963 extended Hückel theory to the treatment of the sigma systems of hydrocarbons. In EHT the elements of the $\mathcal{H}$ matrix are obtained by using atomic ionization potentials for α's and the Wolfsberg-Helmholtz approximation (Equation 24 of Chapter 1) for calculating β's. K is usually set equal to 1.75 and Slater Type Orbitals (STO's)[1] are used to compute the required overlap integrals, which are calculated between all atomic orbitals, even between those on atoms that are not nearest neighbors. Overlap is included throughout the calculation; so that when two orbitals mix, the upper emerges destabilized by a greater amount than the lower is stabilized.

Extended Hückel Theory (EHT) suffers from all the disadvantages of Hückel theory, plus a few more. Because the off-diagonal matrix elements are all attractive and nuclear repulsion is ignored, it might appear that optimization of geometry for an array of atoms would eventually lead to its collapse into a molecule with all bond lengths equal to zero—a sort of super-atom. This is indeed what happens to H_2 when it is treated by EHT; but atoms which bear p orbitals do show energy minima before collapse, because of the node at the nucleus in p orbitals.

Problem 10.1. Show with a drawing of the orbitals involved why a C—H bond, in contrast to an H—H bond, does not tend to a bond length of zero in an EHT calculation.

Although EHT calculations cannot be expected to provide accurate geometries or energies, they can be expected to make qualitatively correct predictions about relative stabilities of similar systems, *provided* that the only important differences in the actual molecules come from the one-electron operators in the Hamiltonian. EHT calculations should, therefore,

---

[1] STO's have the general form for their radial components

$$\phi \sim r^{n-1}e^{-\zeta r}$$

where n is the principal quantum number (1, 2, etc.), and ζ is a parameter which determines the size of an orbital. A standard set of ζ's may be chosen (for instance, by the prescription given by Slater, which is found in most elementary texts on quantum chemistry), or the ζ's may be treated as parameters to be optimized.

be very good at differentiating, for instance, between allowed and forbidden transition states; and, in fact, EHT calculations were used to verify the first theoretical conclusions of Woodward and Hoffmann regarding electrocyclic reactions. However, EHT calculations cannot be expected to give an adequate description of forbidden transition states, or, more generally, any biradical species, because the Hückel Hamiltonian contains no terms by which the mixing of the two configurations, required for the wavefunction of a singlet biradical, can occur. What EHT can do is predict which of two molecular orbitals, both of which appear to be approximately nonbonding, lies lowest and hence will be "more occupied" in the correct many-electron wavefunction. The results of EHT calculations appear to have led to Hoffmann's discovery that through-bond interaction of two "nonbonded" AO's could be substantially greater than that directly through-space.

Despite its many drawbacks, several advantages of the EHT method over more sophisticated ones, which include the electron repulsion operator in the Hamiltonian, should be noted. First, EHT calculations are easy to carry out. Extensive parameter optimization is not required; the chemist need only supply to the program the coordinates of the atoms in the system of interest. Since an EHT program takes little time to complete a calculation, even on a relatively large molecule, EHT calculations are relatively economical of computer time and costs.

Second, the results of EHT calculations are relatively easy to interpret. Because EHT writes total energies as sums of orbital energies, from the results of an EHT calculation it is easy to determine which orbitals are responsible for a particular effect manifested in the total energy. In a calculation which explicitly includes electron repulsion, the total energy is not a sum of orbital energies; consequently, such a determination is much less straightforward. In addition, in an EHT comparison of two similar molecules or of two geometries for the same molecule, energy differences are usually traceable directly to differences in bond orders. Because of the simple relationship between bond orders and energies in EHT calculations, energy differences can often be easily analyzed in terms of differences in bonding between just a few atoms. In calculations which include electron repulsion, because of the more complicated energy expressions, such a simple analysis may be fundamentally correct but more difficult to arrive at. Thus, for ease in finding or confirming a qualitative explanation of a calculated or experimental energy difference, due primarily to a difference in bonding, EHT has a distinct advantage over more sophisticated theoretical methods. Moreover, a hypothesis arrived at and/or verified by an EHT calculation may be further tested by comparing the EHT results with those from a more sophisticated calculation, for instance, by checking that both methods give similar wavefunctions, bond orders, and charge densities. In comparing the results of an EHT calculation with those obtained from one that neglects overlap in order to avoid the n^4 problem in including electron repulsion, it is important to remember that overlap is included in normalizing EHT wavefunctions. Therefore, EHT bond orders and charge densities, calculated from AO coefficients normalized in this way, will be smaller than those obtained from calculations where overlap is ignored in normalization.

The fact that EHT does not neglect the overlap integral is its third advantage over calculations that are in other ways more sophisticated, but which achieve some of their sophistication by neglecting this quantity. Inclusion of the overlap integral is, of course, necessary for correctly describing the net destabilization caused by overlap between two filled orbitals. EHT, for instance, correctly predicts a barrier to rotation in hydrazine, due to the "overlap repulsion" associated with having the nitrogen lone pairs eclipsed, while lone pair repulsion is not correctly described by theoretical methods in which the overlap integral is neglected.

Problem 10.2. Account for the observed barrier to rotation in ethane in terms of overlap repulsion between the C—H bonds on different carbons.[2]

Problem 10.3. In bicyclo[2. 1. 1]hexene the two filled cyclobutane Walsh orbitals that can interact most strongly with the π orbitals of the unsaturated bridge are shown in Figure 10.1. The e orbital is of about the same energy as the ethylene bonding π MO and of higher energy than the b_1 ring orbital. You may assume that the antibonding ring orbitals are of sufficiently high energy that they may be neglected in analyzing the π type interaction between the ring and the bridge. Construct a diagram to analyze the effect of this orbital interaction in bicyclo[2. 1. 1]hexene. Between which two orbitals is the mixing expected to be strongest? What effect do you expect the interaction between ring and bridge to have on the lowest ionization potential of this molecule, compared to that of *cis*-2-butene as a reference?

---

[2] See *J. Amer. Chem. Soc.*, **92**, 3799 (1970) for the answer and *J. Amer. Chem. Soc.* **96**, 3759 (1974) for a fuller discussion.

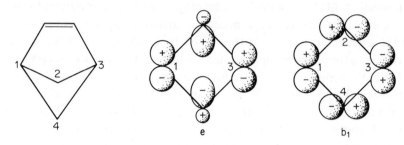

Figure 10.1

Problem 10.4.   An EHT calculation predicts a slightly negative π bond order between the ring and bridge in bicyclo[2. 1. 1]hexene, while that calculated using a method which includes electron repulsion but neglects overlap is found to be positive. Explain. What do you predict to be the actual consequences of the orbital interactions between the ring and bridge for the stability of bicyclo[2. 1. 1]hexene?[3]

Inclusion of Electron
Repulsion—Approximations

We have already seen that it is advantageous to ignore overlap charge distributions in carrying out calculations on π systems in which electron repulsion is included. In an all-valence electron calculation avoidance of the n^4 problem, encountered when differential overlap is not neglected, is even more desirable, since each carbon atom bears not one but four orbitals. Thus, neglect of differential overlap might be said to be roughly $4^4 = 256$ times more desirable in an all-valence than in a π electron calculation; nevertheless, neglect of overlap introduces complications into calculations on all-valence electrons that are not encountered in those dealing only with π systems.

Consider the repulsion between an electron in each of the p orbitals and one in the $1s$ hydrogen orbital, in the geometry shown in part **A** of Figure 10.2. The repulsions would be computed as

$$\gamma_{p_xH} = (p_x p_x | ss)$$
$$\gamma_{p_yH} = (p_y p_y | ss)$$
(1)

[3] See *J. Amer. Chem. Soc.* **95**, 6649 (1973) for the answers to Problems 10.3 and 10.4.

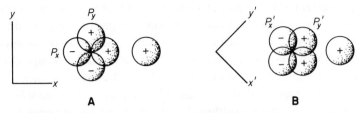

A   **B**

Two Coordinate Systems for Computing Electron Repulsion Integrals Between
$1s$ and $2p$ Orbitals

Figure 10.2

However, there is not a preferred orientation of our coordinate system, since we know the Hamiltonian to be invariant to a rotation of the coordinate axes. Therefore, with the choice of axes shown in **B**, orbitals with the orientation of p_x and p_y in **A** can be written as

$$p_x = \frac{1}{\sqrt{2}}(p'_x + p'_y)$$

$$p_y = \frac{1}{\sqrt{2}}(p'_y - p'_x)$$

(2)

so that the repulsion integrals in (1) become

$$\gamma_{p_zH} = \tfrac{1}{2}(p'_x p'_x + p'_y p'_y + 2p'_x p'_y \,|\, ss)$$

$$\gamma_{p_yH} = \tfrac{1}{2}(p'_x p'_x + p'_y p'_y - 2p'_x p'_y \,|\, ss)$$

(3)

We note that in the primed coordinate system of **B**, the overlap charge distributions given by $p'_x p'_y$ are required to distinguish between the repulsion integrals γ_{p_zH} and γ_{p_yH}. Therefore, if we neglect differential overlap, we find that

$$\gamma_{p_zH} = \tfrac{1}{2}(p'_x p'_x + p'_y p'_y \,|\, ss) = \gamma_{p_yH}$$

(4)

However, (4) must be as valid an expression for these quantities as (1), if we require, as we must, that the results of our calculations cannot depend on the totally arbitrary choice of the orientation of our coordinate system. This implies that the assumption of zero differential overlap, coupled with the requirement of invariance of a calculation to rotation of the coordinate system, demands that all electron repulsion integrals between orbitals on different atoms must be given the same value, regardless of the orientation of the orbitals. Thus, Complete Neglect of Differential Overlap (CNDO) in an all-valence electron calculation entails a rather drastic further approximation, for it requires that p orbitals be treated as if they were spherically symmetric like s orbitals. Although different repulsion parameters could still be used for s and for p orbitals in CNDO calculations, one common value is generally assigned to them, since both must be treated as spherically symmetric functions. Therefore, in a CNDO calculation repulsion integrals between valence orbitals on different atoms depend only on the types of atoms and not on the types of orbitals involved.

Problem 10.5.   Show that when differential overlap is completely neglected, the electron repulsion between an sp^n hybrid pointed at a hydrogen atom is the same as that between the hydrogen atom and an sp^n hybrid pointed away from it. [Hint: Write equations for the two hybrids and the repulsion integrals.] Show that although the overlap integral between an s and a p orbital on the same atom is always zero, the electron density arising from their overlap can give rise to a nonzero repulsion with a hydrogen atom suitably oriented with respect to the p orbital. What must the orientation of the hydrogen atom be for the sp overlap charge density to cause zero net repulsion?

The CNDO Method

Calculations using the CNDO approximations were first carried out by Pople and Segal. In their original calculations a quantity called a penetration integral was retained. It led to a net attraction between two atoms, due to interpenetration of their electron clouds, even when the bond order between them was zero. In addition, the original Pople-Segal expression for the potential between an electron on atom A and the nucleus of A with an effective positive charge of Z_A (Z_A = the number of valence electrons on A) underestimated this attraction in some atoms. Both these defects were remedied in a second version of the CNDO method, appropriately called CNDO/2. The penetration integral was dropped and the one center attraction, U_{rr}, was set equal to

$$U_{rr} = -\tfrac{1}{2}(I_r + \alpha_r) - (Z_A - \tfrac{1}{2})\gamma_{AA} \tag{5}$$

The expression in CNDO/2 for a diagonal $\mathfrak{F}$ matrix element, F_{rr}, becomes

$$F_{rr} = -\tfrac{1}{2}(I_r + \alpha_r) + [(P_{AA} - Z_A) - \tfrac{1}{2}(P_{rr} - 1)]\gamma_{AA} \\ + \sum_{B \neq A} (P_{BB} - Z_B)\gamma_{AB} \tag{6}$$

Note that if all the atoms in a molecule are neutral (so that $P_{AA} = Z_A$ and $P_{BB} = Z_B$ for all B atoms) and if $P_{rr} = 1$, (6) reduces to

$$F_{rr} = -\tfrac{1}{2}(I_r + \alpha_r) \tag{7}$$

which was our definition of α in Chapter 1. Recalling that $\alpha_r = I_r - \gamma_{rr}$, (7) becomes

$$F_{rr} = -I_r + \tfrac{1}{2}\gamma_{rr} \tag{8}$$

which is identical to equation (55) of Chapter 8, the expression for the diagonal $\mathfrak{F}$ matrix element in a π calculation on a neutral AH. It should be noted that in the Pople-Segal parameterization of CNDO/2, $\gamma_{AA} \neq \gamma_{rr}$, since all the electron repulsion integrals are computed theoretically from Slater s orbitals, while γ_{rr} is the semiempirical value of the electron repulsion integral for orbital r, obtained from the experimental energy difference between I_r and α_r. In the Pople-Segal parameterization of CNDO/2, resonance integrals between different orbitals are written

$$\beta_{rs} = \tfrac{1}{2}(\beta_A^0 + \beta_B^0)S_{rs} \tag{9}$$

where the β_A^0 parameters are the same for all orbitals on atom A and are chosen to give the best fit to theoretical calculations. The off-diagonal elements in the CNDO/2 $\mathfrak{F}$ matrix have the same form as those in π electron calculations, namely

$$F_{rs} = \beta_{rs} - \tfrac{1}{2}P_{rs}\gamma_{AB} \tag{10}$$

The Pople-Segal parameterization of CNDO/2 predicts molecular geometries and dipole moments in reasonable agreement with experiment. Alternative parameterizations have been attempted by others in order to

obtain better predictions of total energies, ionization potentials, and spectral transitions.[4] One defect, inherent in the CNDO/2 method itself, that
is irremediable by any change of parameters, is the inability of CNDO/2
to predict singlet–triplet energy differences when the unpaired electrons
occupy different AO's on the same atom. Thus the CNDO/2 method is
unable to deal properly with methylene (CH_2), since CNDO/2 predicts
the same energy for all configurations of the nonbonding electrons in the
sp^n hybrid and the p orbital on carbon. The triplet, of course, lies lowest;
it is the ground state of CH_2. The energy of the triplet differs from that
of the corresponding singlet configuration by twice the exchange integral
$(sp^n, p \mid sp^n, p)$, which at the CNDO/2 level of approximation is set equal
to zero. Moreover, the same sort of integral is responsible for the appearance of unpaired electron density in the sigma system of a π radical, for
it is the source of the energy difference that produces the negative spin
density at the hydrogens in the nodal plane of such a system. Therefore,
the CNDO/2 method cannot be expected to give a complete picture of
the spin distribution in a free radical.

The INDO Method

The most convenient way to remedy this deficiency of the CNDO/2
method is to not ignore integrals of the type $(s, p_x \mid s, p_x)$ and $(p_x, p_y \mid p_x, p_y)$,
provided that the orbitals are *all* on the same atom. Interatomic overlaps
are still ignored, and the resulting method is known as INDO, the acronym
for Intermediate Neglect of Differential Overlap.

Problem 10.6.   Show that in the CNDO method it is impossible to
assign different values to the one-center repulsion integrals $(p_x, p_x \mid p_x, p_x)$
and $(p_x, p_x \mid p_y, p_y)$, where p_x and p_y are on the same atom, while maintaining rotational invariance. [Hint: Examine these integrals in the primed
coordinate system.] Show that these repulsion integrals should be assigned
different values in the INDO formalism. By what quantity do they differ?
Why are integrals like $(p_x, p_x \mid p_x, p_y)$ still zero, even in INDO?

Two additional parameters are required in an INDO calculation
that are not necessary for one using the CNDO methodology. These

---

[4] See the Further Reading section at the end of this chapter for references to recent
reviews.

parameters are the values of $(s, p_i | s, p_i)$ and $(p_i, p_j | p_i, p_j)$, when s, p_i and p_j are all on the same atom; and these parameters may be determined semiempirically from atomic spectral data. Since in all other respects INDO is very similar to CNDO/2, it is not surprising that for closed-shell systems the two methods give very similar results, provided that they are similarly parameterized.[5] The chief motivation for the development of INDO was, of course, its application to open-shell systems; and here, as expected, it performs much better than CNDO/2. For instance, Pople, Beveridge, and Dobosh carried out open-shell, unrestricted (DODS) SCF calculations on methylene using the INDO method. Not only did they find that the triplet was below the singlet in energy and that the singlet was bent, with an HCH angle at the energy minimum of a little over 100°; they also discovered that the triplet had a rather shallow energy minimum around 130°. Significantly (for those who claim that theoreticians publish only those calculations which agree with experiment and consign to the wastebasket the ones that fail to do so), at the time these calculations were published it was generally accepted that the available experimental data showed triplet methylene to be linear or nearly so. In fact, subsequent epr studies by Wasserman and co-workers demonstrated that triplet methylene is bent with an HCH angle of 136°.

Pople and co-workers have also used the INDO method successfully to reproduce theoretically the experimental trends in the epr hyperfine coupling constants of free radicals. Theory predicts a linear relation between the hyperfine coupling constant a_A for an atom A and the unpaired spin density in the valence s orbital of A. Pople calculated the spin density in the valence s orbital of ^1H, ^{13}C, ^{14}N, ^{17}O, and ^{19}F in a number of free radicals and plotted these spin densities against the observed hyperfine coupling constants. In the case of ^1H, ^{13}C, and ^{19}F, linear plots with correlation coefficients on the order of 0.90 were obtained. The ability of INDO to reproduce the McConnell relationship, Equation (14) of Chapter 5, for carbon π radicals was also tested by plotting the calculated π spin density at each carbon against the calculated spin density in the $1s$ orbital of the hydrogen bonded to that carbon. The validity of the McConnell equation demands that such a plot be linear, as indeed it was found to be. From the slope a predicted average value of $Q = 22$ gauss was deduced.

---

[5] Dewar and co-workers have published a number of papers in which they use a Modified INDO (MINDO) method, parameterized to predict the heats of formation of molecules and transition states of interest to the organic chemist. See the Further Reading section for a leading reference.

Trends in nmr coupling constants have also been adequately reproduced by calculations carried out by Pople and co-workers at the INDO level of approximation.

NDDO and Other Approximate Methods

At the next level of approximation only Neglect of Diatomic Differential Overlap (NDDO) is maintained. Integrals like $(s_A, p_{iA} | s_B, p_{iB})$ and $(p_{iA}, p_{jA} | p_{kB}, p_{lB})$ are not set equal to zero, so repulsions between different orbitals on different atoms can depend on their relative orientations without necessarily violating the demands of rotational invariancy. Theoretically calculated repulsion integrals in an NDDO scheme will, of course, maintain rotational invariancy; but if semiempirical values are used, care must be taken to ensure that rotational invariancy is not violated. Considerably more work is required to parameterize and carry out NDDO than CNDO or INDO calculations, and this may account for the relatively small number that have appeared in the literature.

Semiempirical calculations which employ Neglect of Nonbonded Differential Overlap (NNDO) have recently been reported for π systems, but their extension to sigma systems should be feasible. In NNDO the overlap integral between bonded atoms is maintained, so that overlap repulsions are correctly predicted. Thus, one of the better features of the EHT method would be included in a calculation that also takes account of the electron repulsion operator. Because of the explicit inclusion of overlap, Baird[6] found in the first application of the NNDO method to π systems that spectra could be satisfactorily predicted with the same parameters that were used to obtain accurate heats of formation. The n^4 problem is partially avoided in NNDO by the neglect of the very small overlap of orbitals on atoms that are more than 1.8 Å distant and consequently not bonded to each other. In Baird's calculations on π systems the differential overlap between two bonded atoms is also repartitioned using the Mulliken approximation, to further facilitate the computation of electron repulsion integrals. A similar repartitioning, consistent with the assumptions made about one-center differential overlap and with rotational invariancy, would make possible a series of all-valence electron calculations in which the overlap integral was maintained, as in EHT, but in which electron

[6] *Mol. Phys.*, **18**, 29 (1970).

repulsion was treated at the level of approximation of CNDO, INDO, or NDDO. Such a repartitioning of all interatomic overlaps has in fact, already been carried out by Jesaitis in an all-valence electron calculation employing the CNDO approximation for one-center overlaps.[7]

The above does not constitute, by any means, a complete catalog of the semiempirical methods that have been developed for all-valence electron calculations;[8] and undoubtedly, more will appear in the future. We now, however, turn our attention from schemes for semiempirical calculations on valence electrons to the purely theoretical, but nevertheless approximate, ab initio treatment of all electrons.

Ab Initio LCAO Calculations

Pople's name is again that which is most closely associated with recent ab initio calculations of interest to the organic chemist. In his calculations the LCAO approximation is made, and STO's have been the most extensively used to represent atomic orbitals.[9] The elements of the $\mathfrak{F}$ matrix are all computed purely theoretically, using the correct mathematical expressions for the kinetic and potential energies involved and without neglecting any integrals. In order to make easier the mathematics of the computation of the many two-electron integrals required, each STO is actually replaced by a linear combination of gaussian functions, (e^{-ar^2}), which give the best fit to the optimized Slater type orbital. These calculations are called by Pople STO-NG, where N is the number (usually 3 or 4) of gaussian functions used to fit an STO. As we noted in our discussion of H_2 in Chapter 1, calculations on molecules in which the STO exponent is optimized give values of ζ that are larger (so that the orbital is more contracted) than the free atom values suggested by Slater. Therefore, Pople uses a standard set of ζ's which differ from Slater's atomic values.

All of the calculations that we have discussed thus far are done with minimal basis sets—that is, one function is used to represent each atomic orbital, with one coefficient-to-be-determined assigned to it. In order to

---

[7] The reference is given at the end of this chapter.

[8] See, for instance, T. A. Halgren and W. N. Lipscomb, *J. Chem. Phys.* **58**, 1569 (1973) for the Partial Retention of Diatomic Differential Overlap (PRDDO) method.

[9] Least Energy Minimized AO's (LEMAO's) have also been employed. These are AO's which are designed to minimize the calculated energies of isolated atoms.

obtain a better MO wavefunction, an extended basis set can be used, in which more than one function, each with its own coefficient-to-be-determined, is used in place of an AO in the linear combination for the MO[10]. Pople has made use of extended basis sets in some of his ab initio calculations, using a 4G minimal basis set for inner shell AO's but employing two functions for each valence AO—an inner 3G function and a single gaussian for the outer part of the AO. This allows, for instance, an atom to become somewhat anisotropic in an environment of low symmetry. Pople terms this type of extended basis set (4-31G).

Pople and co-workers have found that STO-3G ab initio calculations give equilibrium geometries and dipole moments in good agreement with experiment. Heats of atomization are expected to be calculated poorly ab initio, because in such a calculation electron repulsion integrals are computed purely theoretically, and no account can be taken of electron correlation through semiempirical parameter adjustment. However, for reactions of the type shown in the following equation, in which the number of bonds of the same formal type (C—H, N—H, C—N, C=O, etc.) does not change, STO-3G gives calculated energies with a mean deviation from experiment of about 6 kcal/mole. The 4-31G basis set does even better in calculating the energies of such reactions, which Pople terms "isodesmic," almost halving the absolute mean error.

$$H_2NCHO + CH_4 \longrightarrow CH_3NH_2' + CH_2O$$

The ability of ab initio calculations to predict with some quantitative success the energies of isodesmic processes probably results from a near cancellation of the correlation energies in bonds of the same formal type on both sides of the equation. Since such cancellation is by no means necessarily guaranteed, the accurate prediction of the relative energies of different molecules by ab initio methods cannot be taken for granted. If the amount of correlation energy—that is, the overestimation of the total energy by an SCF calculation, due to the inability of the SCF wavefunction used to provide for the correlation of the motions of electrons of

---

[10] Many calculations have been published which attempt to determine the importance of d orbital participation in the bonding of elements like phosphorous and sulfur from the difference in the energies computed with and without inclusion of d orbitals. Unless the AO's in the molecule are already well represented by the basis set, excluding the functions for the d orbitals, such calculations will only reflect the difference between a minimum and an extended basis set, without necessarily giving any information on the importance of d orbital participation in the actual molecule.

opposite spin—depends on the conformation of a system, then, of course, ab initio calculations cannot even be expected to provide quantitatively accurate results on the relative total energies of different conformations of the same molecule. A good example of a situation where the amount of correlation energy depends on geometry is a molecule in which a bond is being broken. The correlation energy of the two electrons in a bond is large, especially if a single Slater determinantal wavefunction is used. Even if some CI is included, ab initio calculations still tend to overestimate the electron repulsion between the two electrons in a bond. In contrast, when a bond is broken, the linear combination of two Slater determinants, necessary to represent the resulting diradical, correlates the motions of the two electrons very well, so that at the diradical stage the correlation energy is small.

Another problem with energy differences computed by ab initio methods arises from the fact that the limited basis sets, used for organic molecules of interest, will not give even the best possible SCF wavefunctions. It is quite likely that these approximate wavefunctions may represent one class of molecules better than another, especially if the wavefunctions are derived from a rather inflexible minimum basis set. For instance, the wavefunctions for strained systems may be much less well represented than those for unstrained molecules. Indeed, this is probably the case with Pople's STO-3G and 4-31G wavefunctions; for the energies of isodesmic reactions involving small rings that are calculated from these basis sets tend to overestimate strain energies. Significantly, the extended basis exaggerates strain energies considerably less than the minimal one.[11]

Summary

We have seen that there are three distinct classes of theoretical methods for treating all-valence electrons. Most approximate is the semiempirical EHT method, which takes no account of electron repulsion. Semiempirical methods that include the two-electron operator in the Hamiltonian are approximate in varying degree, depending on how they deal with differential overlap. Least approximate, with respect to retaining both repulsion integrals and overlap, are ab initio methods.

[11] In contrast, most semiempirical methods of the _NDO type have a tendency to underestimate strain energies, with the result that cyclic arrays of atoms tend to be greatly favored over acyclic arrays by such calculations.

Which type of method the organic chemist chooses for a particular problem depends on a number of variables, including the computing facilities that are available to him and the computing costs that he is willing to pay. The most important variable, however, is what the chemist hopes to accomplish with a calculation. Depending on his goal, the chemist will of course, want to choose the method that holds the most promise of helping him achieve it. For instance, as we have seen, it is pointless to carry out CNDO/2 calculations on carbenes, since this method is incapable of distinguishing singlets from triplets. On the other hand, carrying out an ab initio calculation just to find out how the orbitals of reactant are transformed into those of the product in the ring opening of cyclohexadiene is like using a cannon to kill a fly.

In order to decide intelligently whether a given method is suitable for a particular theoretical investigation, it is obvious that the chemist must be familiar with the approximations inherent in and the consequent limitations of that method. The chemical literature is replete with EHT treatments of ions and ionic compounds, calculations on biradicals using single Slater determinantal wavefunctions, and similar misapplications of theory. Knowledge sufficient to judge the applicability of a method to a particular problem is not only necessary for the organic chemist intending to do calculations of his own, but also for the organic chemist who just wishes to be able to read critically the burgeoning literature of theoretical-organic chemistry.

Problem 10.7.   What type of calculation, EHT, _NDO, or ab initio, would be most applicable to the following problems? Explain why.

(a)   Determination of the geometry of lowest energy for the pentavalent carbocation CH_5^+.

(b)   Verification that the preference for a chair transition state in the Cope rearrangement is due to a negative bond order (antibonding) between two formally nonbonded carbons.

(c)   Prediction of the energy of the first allowed electronic transition in a series of hydrocarbons.

Problem 10.8.   While the results of a sophisticated calculation may provide novel information about a chemical system, the system cannot be said to be *understood* any better unless the results of the calculation are amenable to a qualitative explanation. Provide such an explanation in terms of perturbation theory for the following theoretical results.

Figure 10.3

(a)   When X is more electronegative than hydrogen, conformation **B**, shown in Figure 10.3, is found by ab initio calculation to be preferred for the substituted ethyl carbocation, while **A** is preferred for the anion. When X is less electronegative than hydrogen, the reverse conformational preference is found. [Hint: Consider a component of the C—X bond in **A** and the C—H bonds in **B** as overlapping effectively in a π fashion with the π orbital on the adjacent carbon. How will the energy of and the coefficents in the filled bonding and the unfilled antibonding C—X orbital change with the electronegativity of X?][11]

(b)   CNDO/2 calculations show that cyclopropane is better at stabilizing an adjacent carbocation than is ethylene, a theoretical conclusion that confirms abundant experimental evidence regarding this point. [Hint: The highest filled orbitals in cyclopropane are a degenerate pair of Walsh orbitals, which are comprised of carbon p orbitals oriented around the periphery of the ring (as in the b_1 cyclobutane orbital in Figure 10.1), whose energy is only slightly lower than that of the π orbital in ethylene. Use group theory to find these MO's in real form and see which can mix with an empty p orbital on a carbon atom attached to the ring.] Do you predict a conformational dependence for the stabilization by cyclopropane?[12]

(c)   Detailed ab initio calculation [*J. Amer. Chem. Soc.* **96,** 959 (1974)] of the potential surface for the allowed mode of ring opening in the cyclopropyl cation shows this reaction to occur with synchronous bond stretching and methylene rotation. In contrast, the calculated potential surface for the allowed mode of ring opening in the anion shows bond stretching to precede rotation. [Hint: Consider the interactions between AO's that develop as the reaction proceeds toward the product. If the correct explanation still remains unclear, see Problem 3.3.] Do you predict that bond stretching and methylene rotation will be synchronous in

[11] The answer to this problem is contained in *J. Amer. Chem. Soc.,* **94,** 6221 (1972).

[12] The answer to this problem is contained in *J. Amer. Chem. Soc.,* **95,** 8193 (1973).

the allowed mode of ring opening in cyclobutene? Make a similar prediction for 1,3-cyclohexadiene.

FURTHER READING

J. A. POPLE and D. L. BEVERIDGE review the development and applications of semiempirical MO methods in *Approximate Molecular Orbital Theory*, McGraw-Hill, New York, 1970.

G. KLOPMAN and B. O'LEARY, *Fort. Chem. Forsch.*, **15**, 445 (1970) also provide a review of semiempirical MO methods, particularly those that include electron repulsion. Their article features a series of helpful tables which list for each method: the authors responsible for its development and the corresponding literature reference, expressions for the various integrals involved, as well as for the elements of the $\mathfrak{F}$ matrix and the total energy, applications, successes and failures, and even the Quantum Chemical Program Exchange number of the source program.

J.N. MURRELL and A.J. HARGET present the most up-to-date account in book form of *Semiempirical SCF-MO Theory of Molecules*, Wiley-Interscience, New York, 1972.

A paper, by N. BODOR, M. J. S. DEWAR, and D. H. LO, *J. Amer, Chem. Soc.*, **94**, 5303 (1972) describes the latest parameterization of the MINDO method, MINDO/2′, and provides leading references to previous versions, MINDO/1 and MINDO/2, and their applications. Calculations involving yet another parameterization, MINDO/3, have recently been reported in *J. Amer. Chem. Soc.*, **96**, 253 (1974).

R. G. JESAITIS, *J. Amer. Chem. Soc.*, **93**, 3849 (1971) compares the results of calculations where diatomic differential overlap is not neglected but repartitioned, using the Mulliken approximation, with those obtained from CNDO/2. Some failings of the latter method are discussed.

J. A. POPLE, *Accts. Chem. Res.*, **3**, 217 (1970) provides a brief review of molecular orbital methods in organic chemistry with a heavy emphasis on ab initio calculations. Early work of the Pople group on ab initio calculations of the energies of isodesmic reactions is found in

W. J. Hehre, R. Ditchfield, L. Radom, and J. A. Pople, *J. Amer. Chem. Soc.*, **92**, 4796 (1970).

Sigma Molecular Orbital Theory, edited by K. B. Wiberg and O. Sinano-glu, Yale University Press, New Haven, Conn., 1970, is a collection of short articles—reprints of the most important papers in this area as well as many contributions prepared especially for this volume.

Scientific Citations lists all the articles each year which cite as a reference a particular paper. The listings are by author of the paper cited. Articles reporting recent applications of a particular method can be easily located, since such articles will almost certainly cite the original paper reporting the development of that method. In addition to an edition for each past year, volumes of *Scientific Citations* appear during each current year covering each quarter.

The Organic Chemist's Book of Orbitals, W. L. Jorgensen and L. Salem, Academic Press, N. Y., 1973, is just what its name implies: a series of drawings, generated by computer, of all the molecular orbitals in a number of relatively simple molecules of interest to the organic chemist. The first part of the book consists of a clear qualitative explanation of how molecular orbitals are constructed from local bond and group orbitals.

Index

R

S